Marine Engineering English

선박기관 영어

장 세 호 저

preface

초급사관이 처음으로 선박에 승선하면 여러 가지의 많은 어려움과 직면하게 된다. 그중에서도 외국인 사관 또는 부원과 함께 근무하는 경우에 이들과의 일상적인 의사소통 뿐 아니라 업무와 관련된 의사 전달의 어려움이 매우 크며, 그리고 책임 사관으로서 작성해야 하는 영문기관일지 작성과 기관실의 각종 기기 운전, 장비를 위한 영문 취급설명서 해석을 들 수 있다.

이러한 어려움을 극복하기 위해서는 우선적으로 기관실에서 사용하는 관련 용어와 약어 및 각종 명칭을 먼저 파악해야 하며, 기관실에서 일어나는 각종 사고와 업무를 기관일지에 영문으로 작성하기 위해서는 기관실에서 많이 사용되는 각종 표현법과 전공 영어를 숙지 할 필요가 있다.

기관일지는 단순기록에 그치는 것이 아니라 각종 해난 사고가 발생할 경우에는 증거자료로 활용되며, 주기를 비롯한 각종 보조기계들의 운전과 정비를 함에 있어 중요한 기초 자료로 사용되므로 기관일지 작성은 정확하고 구체적으로 표현되어야 한다.

따라서 본 교재는 초급사관으로 승선하여 직면하게 되는 기관일지 작성과 일상적인 의사소통 및 직무와 관련된 영어화화에 조금이나마 도움을 주고자 하였으며, 각종 해기 관련 시험을 대비한 기관영어 문제도 함께 수록하였다. 끝으로 이 교재를 편집하는데 많은 도움을 주신 현경은 교수님께 깊은 감사를 드린다.

2018. 10. 31

저자 씀

contents

01

Terms Used in the Engine Department

기관부 관련 용어

1.1 Common Abbreviation in Engine Room (기관실 상용 약자)

약 자	영 문	뜻
C/E	Chief Engineer	기관장
1/E	First Engineer	1등 기관사
2/E	Second Engineer	2등 기관사
3/E	Third Engineer	3등 기관사
S/B	Stand By	스탠바이, 준비완료
D/H	Dead Slow Ahead	극미속 전진
S/H	Slow Ahead	미속 전진
H/H	Half Ahead	반속 전진
F/H	Full Ahead	전속 전진
D/S	Dead Slow Astern	극미속 후진
S/S	Slow Astern	미속 후진
H/S	Half Astern	반속 후진
F/S	Full Astern	전속 후진
ACC	Automatic Combustion Control	자동 연소 제어
AVR	Automatic Voltage Regulator	자동 전압 조정기
AC	Alternating Current	교류
BHP	Brake Horse Power	제동마력
BDC	Bottom Dead Center	하사점
BLR	Boiler	보일러
BWT	Ballast Water Tank	밸러스트 탱크
C/R	Control Room	제어실
DBT	Double Bottom Tank	이중저(二重底) 탱크
D/G	Diesel Generator	디젤 발전기
D.O	Diesel Oil	디젤유(벙커 A)
DC	Direct Current	직류

약 자	영 문	뜻
DPA	Differential Pressure Alarm	차압 경보
DPIA	Differential Pressure Alarm with Indicating	차압 경보 및 지시
DPS	Differential Pressure Switch	차압 스위치
DPT	Differential Pressure Transmitter	차압 전송기
EHP	Effective Horse Power	유효 마력
E/G	Emergency Generator	비상 발전기
E/R	Engine Room	기관실
F.O	Fuel Oil	연료유
F.O.T	Fuel Oil Tank	연료유 탱크
F.W	Fresh Water	청수(민물)
F.W.G	Fresh Water Generator	조수기
F.W.T	Fresh Water Tank	청수 탱크
FLS	Float Level Switch	레벨 스위치
FS	Flow Switch	유량 스위치
G/E	Generator Engine	발전기
H.F.O	Heavy Fuel Oil	중유(벙커 C)
HTR	Heater	가열기
IHP	Indicated Horse Power	지시 마력
LNG	Liquefied Natural Gas	액화 천연 가스
LPG	Liquefied Petroleum Gas	액화 석유 가스
L.O	Lubricating Oil	윤활유
LAH	High Level Alarm	고수위 경보
LAL	Low Level Alarm	저수위 경보
LC	Level Controller	레벨 제어기
LI	Level Indicator	레벨 지시기
LS	Limit Switch	리미트 스위치
LT	Level Transmitter	레벨 전송기
M.D.O	Marine Diesel Oil	선박용 디젤유

약 자	영 문	뜻
M.S	Motor Ship	동력선
M/E	Main Engine	주기관
MO	Man Zero	무인 기관실
OMA	Oil Mist Alarm	유분 경보
OMD	Oil Mist Detector	유분 검출기
PP, P/P	Pump	펌프
PAH	High Pressure Alarm	고압 경보
PAL	Low Pressure Alarm	저압 경보
PC	Pressure Controller	압력 제어기
PIAL	Low Pressure Alarm with Indicating	저압 경보 및 지시
PI	Pressure Indicator	압력 지시기
PPM	Parts Per Million	백만분의 1, 백만분율
PS	Pressure Switch	압력 스위치
PT	Pressure Transmitter	압력 전송기
R/up	Rung-up	정상 항행
RB	Resistance Bulb	저항 전구
RPM	Revolutions Per Minute	분당 회전수
SHP	Shaft Horse Power	축마력
S.W	Sea Water	해수
S/G	Steering Gear	조타기
SAH	High Salinity Alarm	고염도 경보
SE	Salinity Sensor	염도 센서
SHA	Shut Down Alarm	운전 정지 경보
SI	Salinity Indicator	염도 지시기
SLA	Slow Down Alarm	엔진 감속 경보
T/C	Turbo Charger	과급기
TAH	High Temperature Alarm	고온 경보
TAL	Low Temperature Alarm	저온 경보

약 자	영 문	뜻
TIAH	High Temperature Alarm with Indicating	고온 경보 및 지시
TC	Temperature Controller	온도 제어기
T.D.C	Top Dead Center	상사점
THC	Thermocouple	열전대
TIAL	Low Temperature Alarm with Indicating	저온 경보 및 지시
TI	Temperature Indicator	온도 지시기
TS	Temperature Switch	온도 스위치
VAH	High Viscosity Alarm	고점도 경보
VAL	Low Viscosity Alarm	저점도 경보
VC	Viscosity Controller	점도 제어기
VI	Viscosity Indicator	점도 지시기
VT	Viscosity Transmitter	점도 전송기

기본어휘

어 휘	뜻	어 휘	뜻
ahead	앞으로	astern	뒤로
bottom	밑바닥	ballast	밸러스트
bulb	구형의 것(전구 등)	chief	장(長)
combustion	연소	controller	제어기
control	제어, 조절	current	전류
couple	한 쌍, 짝	double	이중의
detector	검출기	differential	차이
D/E(Dynamo Engine)	발전용 엔진	effective	유효한
emergency	비상	flow	흐름, 유동, 유량
generator	발전기	horse power	마력
liquefied	액화시킨	lubricating	윤활(유를 바른)
million	100만	marine	바다의, 해양의
mist	안개	regulator	조정기
rung	ring의 과거	resistance	저항
revolution	회전	steering	조타, 조종
shut down	운전 정지	slow down	감속
salinity	염분, 염도	S/G(Shaft Generator)	축 발전기
thermo	열의	T/G(Turbine Generator)	터빈발전기
temperature	온도	transmitter	전송기
voltage	전압	viscosity	점도

1.2 Kinds of Machinery (각종 기기)

1. 기기

Air Compressor	공기 압축기	Heater	가열기
Air Conditioner	공기 조화 장치	Heat Exchanger	열교환기
Boiler, BLR	보일러	Incinerator	소각기
Calorifier	온수기	Main Engine	주기관
Condenser	응축기	Oily Water Separator	유수 분리기
Cooler	냉각기	Propeller	프로펠러
Economizer	배기가스열교환기	Purifier	청정기
Emergency Generator	비상 발전기	Refrigerator Machine	냉동기
Fresh water Generator	조수기	Steering Gear	조타기
Generator Engine	발전기	Turbocharger	과급기

기본어휘

어휘	뜻	어휘	뜻
compressor	압축기	oily	기름의
oily water	유수[1)]	separator	분리기
purify	깨끗하게 하다	refrigerate	냉각하다
refrigerator	냉장고		

1) 기름 성분이 섞여 있는 물

1.3 Kinds of Tank (각종 탱크)

Bilge Tk	빌지 탱크	F.W Exp. Tk for M/E	주기용 청수 팽창 탱크
Cascade Tk	보일러 응축수 탱크	F.W Tk	청수 탱크
Daily Tk	상용 탱크	H.F.O Tk	벙커C 수급 탱크
D.O Service Tk	디젤유 서비스 탱크	L.O Settling Tk	윤활유 침전 탱크
D.O Settling Tk	디젤유 침전 탱크	Main L.O Sump Tk	주기 윤활유 섬프 탱크
D.O Tk	디젤유 수급 탱크	Oily Bilge Tk	폐유수 탱크
F.O Service Tk	중유 서비스 탱크	S/T L.O Tk	선미관 윤활유 탱크
F.O Settling Tk	중유 침전 탱크	Waste Oil Tk	폐유 탱크

기본어휘

어휘	뜻	어휘	뜻
bilge	더러운 물[2)]	Tk(tank)	탱크
exp.(expansion)	팽창	Cascade Tk[3)]	보일러 응축수 탱크
settle	침전시키다	S/T(stern tube)	선미관
waste	쓸모없는		

2) 기관실 바닥에 고여 있는 더러운 물
3) 복수된 응축수에 기름의 함유 여부 등을 관찰하기 위해 설치한 탱크

1.4 Kinds of Pump (각종 펌프)

AUX C.S.W PP	보조 냉각 해수 펌프
BILGE PP	빌지 펌프
BLR FEED WATER PP	보일러 급수 펌프
BLR CIRC. PP	보일러수 순환 펌프
CAMSHAFT L.O PP for M/E	주기관용 캠샤프트 윤활유 펌프
CROSSHEAD L.O PP for M/E	주기관용 크로스헤드 윤활유 펌프
C.S.W PP for AIR COND	에어컨용 냉각 해수 펌프
CYL OIL TRANS PP for M/E	주기관용 실린더 오일 이송 펌프
D.O TRANS PP	벙커 A 이송 펌프
EJECTOR PP for F.W GEN.	조수기용 이젝터 펌프
FIRE & G.S PP	소화 및 잡용수 펌프
F.O CIRC. PP for G/E	발전기용 연료유 순환 펌프
F.O SUPPLY PP for M/E	주기관용 연료유 공급 펌프
F.O TRANS PP	연료유 이송 펌프
F.W HYD. PP	청수 가압 펌프
HOT WATER CIRC. PP	고온수 순환 펌프
JACKET C.F.W PP for M/E	주기관용 재킷 냉각 청수 펌프
L.O TRANS PP	윤활유 이송 펌프
MAIN C.S.W PP	주 냉각 해수 펌프
MAIN L.O PP	주 윤활유 펌프
PURI. FEED PP	청정기 공급수 펌프
SLUDGE PP	슬러지 펌프
STERN TUBE L.O PP	선미관 윤활유 펌프

기본어휘

어휘	뜻	어휘	뜻
AUX[4]	보조의	C.S.W[5]	냉각 해수
PP(pump)	펌프	CIRC.[6]	순환하는
TRANS[7]	이송하다	G.S[8]	잡용
SUPPLY	공급하다	HYD.[9]	수압의, 수력의
C.F.W[10]	냉각 청수	PURI.[11]	청정기

4) auxiliary
5) cooling sea water
6) circulating
7) transfer
8) general service
9) hydraulic
10) cooling fresh water
11) purifier

■ EXERCISE

약 자	뜻	약 자	뜻
C/E		EHP	
1/E		E/G	
2/E		E/R	
3/E		F.O	
S/B		F.O.T	
D/H		F.W	
S/H		F.W.G	
H/H		F.W	
F/H		FLS	
D/S		FS	
S/S		G/E	
H/S		H.F.O	
F/S		HTR	
A.C.C		IHP	
A.V.R		LNG	
AC		LPG	
BHP		L.O	
BDC		LAH	
BLR		LAL	
B.W.T		LC	
C/R		LI	
D.B.T		LS	
D/G		LT	
D.O		M.D.O	
DC		M.S.	
DPA		M/E	
DPIA		MO	

약 자	뜻	약 자	뜻
DPS		OMA	
DPT		OMD	
PP, P/P		VC	
PAH		VI	
PAL		VT	
PC		Air Compressor	
PIAL		Air Conditioner	
PI		Boiler, BLR	
PPM		Calorifier	
PS		Condenser	
PT		Cooler	
R/up		Economizer	
RB		Emergency Generator	
RPM		Fresh water Generator	
SHP		Generator Engine	
S.W		Heater	
S/G		Heat Exchanger	
SAH		Incinerator	
SE		Main Engine	
SHA		Oily Water Separator	
SI		Propeller	
SLA		Purifier	
T/C		Refrigerator Machine	
PP, P/P		VC	

약 자	뜻	약 자	뜻
PAH		VI	
PAL		VT	
PC		Air Compressor	
PIAL		Air Conditioner	
PI		Boiler, BLR	
PPM		Calorifier	
PS		Condenser	
PT		Cooler	
R/up		Economizer	
RB		Emergency Generator	
RPM		Fresh water Generator	
SHP		Generator Engine	
S.W		Heater	
S/G		Heat Exchanger	
SAH		Incinerator	
SE		Main Engine	
SHA		Oily Water Separator	
SI		Propeller	
SLA		Purifier	
T/C		Refrigerator Machine	
TAH		Steering Gear	
TAL		Turbocharger	
TIAH		Bilge Tk	
TC		Cascade Tk	

약 자	뜻	약 자	뜻
TDC		Daily Tk	
THC		D.O Service Tk	
TIAL		D.O Settling Tk	
TI		D.O Tk	
TS		F.O Service Tk	
V, V/V		F.O Settling Tk	
VAH		F.W Exp. Tk for M/E	
VAL		F.W Tk	
H.F.O Tk		FIRE & G.S PP	
L.O Settling Tk		F.O CIRC. PP for G/E	
Main L.O Sump Tk		F.O CIRC. PP for M/E	
Oily Bilge Tk		F.O SUPPLY PP for G/E	
S/T L.O Tk		F.O SUPPLY PP for M/E	
Waste Oil Tk		F.O TRANS. PP	
AUX C.S.W PP		F.W HYD. PP	
BILGE PP		HOT WATER CIRC. PP	
BLR FEED WATER PP		JACKET C.F.W PP for M/E	
BLR CIRC. PP		L.O TRANS. PP	
CAMSHAFT L.O PP for M/E		MAIN C.S.W PP	

약 자	뜻	약 자	뜻
CROSSHEAD L.O PP for M/E		MAIN L.O PP	
C.S.W PP for AIR COND		PURI. FEED PP	
CYL OIL TRANS. PP for M/E		SLUDGE PP	
D.O TRANS. PP		STERN TUBE L.O PP	
EJECTOR PP for F.W GEN.			

02

Engine Log Book
기관일지

2.1 Kinds of Engine Log Book

2.2 Entries Engine Log Book

2.3 Examples of Entries in Remark Columns

2.1 Kinds of Engine Log Book (기관일지 종류)

기관일지(Engine Log Book)는 기관실에 있는 각종 기기의 상태 및 작업 내용 등을 기록하는 서식으로서 엔진을 안전하게 운전하고 효율적으로 관리하는 데 필요하며, 또한 해난 사고가 발생했을 경우 공법 및 사법상 중요한 증거 자료가 되므로 정확히 작성하고 관리해야 하며, 화재나 충돌 등 해난 발생 시에는 비상 반출해야 하는 것을 잊어서는 안 된다.

기관사가 기재하고 관리하는 기관부 일지에는 기관장 일지(Chief Engineer's Log Book), 기관 당직 일지(Engine Room Log Book), 기관 적요 일지(Abstract Engine Log Book) 등이 있다.

1. Chief Engineer's Log Book(기관장 일지)

기관장 일지는 기관장이 기관 당직 일지로부터 필요한 사항을 옮겨 기록하고, 당직 기관사와 기관장이 서명하여 선내에 보존하는 가장 중요한 기관부 일지로서 기관장이 관리한다.

2. Engine Room Log Book(기관 당직 일지)

기관 당직 일지는 항상 기관실에 비치해 놓고 당직 기관사가 근무 중 있었던 기관실 각종 기기의 운전 상황 및 작업 내용을 기록하는 것으로서, 대개 기관장 일지와 양식이 동일하므로 편의상 기관장 일지로써 기관 당직 일지를 대신하는 경우가 많다.

3. Abstract Engine Log Book(기관 적요(요약) 일지)

기관 적요 일지는 매 항차 종료시에 기관당직 일지의 요점을 한 장의 소정의 서식에 요약하여 본선의 엔진 상태나 유류 소모량 등을 한 눈에 파악할 수 있도록 정리해 놓은 일지이다. 이것은 선박의 운항 실태를 파악할 수 있는 중요한 자료로서 1부는 본선에 보관하고, 1부는 선주나 용선자에게 제출한다.

2.2 Entries of Engine Log Book (기관 일지 기재 내용)

1. 보관과 취급

① 기관일지는 기관장이 보관한다.

② 기관실의 적당한 장소에 비치하고 당직 중의 필요사항을 당직 기관사는 빠짐없이 기록한다.

③ 기재사항은 사후에 즉시 사실 그대로 쓰는 것이므로 일단 기록하고 나면 변경하지 못한다. 즉, 일지의 잘못 쓴 부분은 흑선 또는 적선을 긋고 당직 사관의 서명 후 아래 또는 부근에 다시 써서 최초의 기록을 읽을 수 있도록 한다.

2. 기재상의 일반적인 주의 사항

① 사실을 충실히 기재한다.

② 기사는 가급적 간단명료하게 쓰고, 애매한 표현은 피한다.

③ 일어난 일을 사후에 기록하는 것이므로 문체는 원칙적으로 과거형으로 쓴다.

④ 장기간 보존해야 하므로 펜으로 기입하여야 하며, 연필로 기입해서는 안된다.

기관일지의 양식과 기재 내용은 회사 또는 엔진의 종류에 따라 약간씩 다르나, 기입하는 요령은 대개 비슷하다. 기관 당직 일지(Engine Room Log Book)는 보통 기관 제어실(C/R)에 기관장이 지정한 그 장소에 항상 보관하여야 하며, 당직 기관사가 근무 중에 메모해 둔 사항들을 기초로 하여 당직 근무를 미치기 직전에 기재한다. 이 일지는 크게 Principal Column(본란)과 Remark Column(기사란)으로 구분되어 있다.

3. 본란

일지의 본란은 주기관 및 중요 보조 기기의 운전과 관련된 사항들을 기재하는 곳으로 정해진 시각에 정확히 기록하여야 한다. 엔진 회전수, 냉각수 및 윤활유의 입·출구 온도와 압력, 흡기와 배기 온도, 유류 소모량 및 기타 주요 부분의 온도와 압력 등을 숫자로 기입한다. 또, 매일의 정오 통계란에는 항해 시간, 평균 회전수, 항해 거리, 평균 속력, 슬립 등을 계산하여 기입한다.

4. 기사란

기사란(Remarks)에는 본란에 기입하는 것 이외에 당직 근무 중 발생한 주요 사항에 대해 그 시각, 상황 및 처리 개요 등을 순서대로 기재한다. 그리고 일지에 기재하는 영문은 문장이 간결하고도 내용이 명확하여야 하므로, 관례적으로 사용하고 있는 표현 방법을 익혀 활용하도록 한다.

① 시간은 4단위 숫자로 기입한다. 즉, 오전 9시25분은 0925, 오후 3시45분은 1545 등과 같이 쓴다.

② 그 날의 작업 상황을 정리해서 기록한다.

5. 기관일지 기재 항목

(1) 첫 페이지에 기재하는 항목

주기관(Main Engine)	
종류(Type)	제조 년월(When Built)
제조 장소(Where Built)	제조자(Builder)
추진기의 종류(Kind of Propeller)	추진기의 피치(Pitch of Propeller)
지시마력(Indicated Horse Power)	연속 최대 마력(Nominal Horse Power)

주보일러(Main Boiler)	
종류(Kind)	재료(Material)
개수(Number)	제조 년월(When Built)
제조 장소(Where Built)	최대 기압(Loaded Pressure)

보조보일러(Donkey Boiler)	
종류(Kind)	개수(Number)
최대 기압(Loaded Pressure)	

(2) 일지에 기재하는 항목

선명(Name of Vessel)
회사명(Name of Company)
항차(Voyage No.)
운항지(From... To...)
주기관 작동시간(Hours of Engine Motion)
적산 회전계 회전수(Revolution on Counter)
평균 매분 회전수(Mean Revolution per Minute)
매시 평균 속력(Average Speed per Hour)
측정기 항주거리(Distance Run, Log)
실측 항주거리(Distance Run, Observation)
추진기 항주거리(Distance Run, Screw propeller)
실각율(Slip Percentage)
실린더오일 소비량(Cylinder Oil Consumption)
연료유 소비량(Fuel Oil Consumption in Watch)
연소기내 연료유압력(Fuel oil Pressure in Burner)
연료유(Fuel oil)
실린더 오일(Cylinder Oil)

6. 영문기관일지 작성 특성

기관일지의 기사란에 기재하는 영문은 문장이 간결하면서도 의미가 명확해야 한다. 그래서 기관실에서 관례적으로 사용하는 문체나 용어를 일반 영문과 비교해 보면, 다음과 같은 몇 가지 독특한 점이 있다.

① 문장의 뜻이 변하지 않는 한 주어가 본선 자체이거나 본선 선원일 때는 이를 생략한다.

(a) Our ship left Busan for Incheon. → Left Busan for Incheon.

(b) We arrived at Incheon. → Arrived at Incheon.

② Be 동사는 의미가 불명확하게 되지 않는 한 생략하며, 날씨를 표현할 때는 주어와 동사 둘 다 생략한다.

(a) The sea was calm. → Sea calm.

(b) Visibility was good. → Visibility good.

(c) Engine was stopped. → Engine stopped.

③ 지명, 선박, 천체 및 기타 일반 명사 등에 붙는 정관사는 생략한다.

(a) the Pacific Ocean → Pacific Ocean

(b) the M/S 'King Star' → M/S 'King Star'

(c) the Main Engine → Main Engine

④ 문장의 요점이 되는 단어 이외에는 생략한다.

(a) No smoking is allowed. → No smoking.

(b) We had fine weather. → Fine weather.

(c) Safety is the first. → Safety first.

(d) Keep your hands off. → Hands off.

(e) No admittance is allowed. → No admittance.

(f) Rung S/B engine. → S/B engine.

⑤ 문장의 시제는 과거형으로 한다.

(a) Received 1000 M/T of H.F.O.

(b) Left Busan for Incheon.

⑥ 본선의 대명사는 3인칭 여성 단수로 한다.

(a) Mr. K left her.

(b) Mr. H joined her as 2/E.

⑦ 약자나 기호를 많이 사용한다.

(a) Finished with engine. → F.W.E

(b) Engine Room → E/R

(c) Steam ship → S.S

(d) Motor ship → M.S

(e) First engineer → 1/E

(f) and → &

2.3 Examples of Entries in Remark Columns
(기사란의 기재 예문)

■ 입 · 출항에 많이 사용되는 용어

약 자	영 문	뜻
S/B	Stand By	스탠바이, 준비완료
F/H	Full Ahead	전속 전진
H/H	Half Ahead	반속 전진
S/H	Slow Ahead	미속 전진
D/H	Dead Slow Ahead	극미속 전진
F/S	Full Astern	전속 후진
H/S	Half Astern	반속 후진
S/S	Slow Astern	미속 후진
D/S	Dead Slow Astern	극미속 후진
	Ring(Rung) up engine	전속 전진
	Finished with Engine	기관사용 종료
	Shifting berth, S/B	묘박지 변경 스탠바이
	Entering the dock, S/B	도킹 스텐바이

■ 기관일지 작성 때 자주 사용되는 표현

예 문	표 현
~에(로부터) ~을 시작했다.	Started ~ at(from) ~.
온도를 ~℃로 유지했다.	Kept the temperature at ~℃.
기관의 ~을 점검했다.	Examined(Inspected) ~ of engine.
~으로 ~을 채웠다.	Filled up with ~ into ~.
~을 떼어냈다.	Removed(Took off) ~.
~을 복구했다.	Replaced(Refitted, Closed up) ~.
~을 ~시에 시운전했다.	Tried ~ at~.
~을 하기 전에 검사했다.	Examined(Inspected) ~ before ~.

예 문	표 현
~하기 위하여 스탠바이	S/B for~.
~에 투묘했다.	Anchored at ~.
~을 ~으로 변경했다.	Exchanged ~ for ~.
~을 개방하고 ~을 들어냈다.	Opened out ~ & took out ~.
~을 검사하고 소제했다.	Examined & cleaned ~.
~을 확실히 죄었다.	Secured ~ in good condition.
~을 개방검사하고 고장 난 ~을 신품과 교환했다.	Overhauled ~ & renewed defective ~ of them.
압력 ~kg/㎠으로 조정했다.	Adjusted pressure at ~ kg/㎠.
~의 파손부분을 수리했다.	Repaired broken part of ~.
~을 다듬었다.	Skimmed up ~.
~의 하나가 파손된 것을 발견했다.	Found one of ~ broken.
~에 ~을 채웠다.	Put(Filled up) ~ into ~.
~을 예비품과 교체했다.	Replaced ~ with a spare one.
A와 B를 교환했다.	Exchanged A for B.
~을 시작했다.	Commenced (Started) ~.
~을 ~에 적재했다.	Loaded(Took in) ~ into ~.
~ 으로부터 ~을 수령했다.	Received ~ from ~.
~을 개방하고 내부 상태를 확인했다.	Opened ~ & inspected inside condition.
~을 계측하고 기록했다.	Measured & noted down ~.
~ 한지를 확인했다.	Ascertained if ~.

1. Preparation of Departure(출항 준비)

① Commenced warming-up for M/E at 1200.
(12시부터 주기관을 워밍(난기)하기 시작하였다.)

② Warmed up M/E for 2 hours.
(주기관을 2시간 동안 워밍하였다.)

③ Engaged turning gear and turned M/E for 20 minutes.

(터닝 기어를 맞물려서 주기관을 20분 동안 터닝하였다.)

④ Disengaged turning gear.

(터닝 기어를 뺐다.)

⑤ Started No.1 G/E at 0900 & set Nos. 1 & 2 generators in parallel running.

(오전 9시에 1번 발전기를 기동시켜 2호기와 병렬 운전하였다.)

⑥ Tried M/E and confirmed it in good condition.

(주기관을 시운전한 결과, 상태가 양호함을 확인하였다.)

⑦ Primed up all fuel injection lines of M/E.

(주기관의 모든 연료 분사관 계통을 프라이밍했다.)

⑧ Ran the engine several revolution with starting air alone before trial on F.O.

(연료로 시운전하기 전에 시동공기만으로 수회전 기관을 운전했다.)

2. Departure(출항)

① S/B eng. for leaving Port of Busan.

(부산항을 떠나기 위해 엔진을 준비하였다.)

② Started (or Stopped) M/E.

(주기관을 기동(또는 정지)시켰다.)

③ Changed F.O for M/E from M.D.O to H.F.O.

(주기관의 사용 연료를 M.D.O에서 H.F.O로 바꾸었다.)

④ Changed bunker 'A' to 'C' at 1300.

(오후 1시에 벙커를 'A'에서 'C'로 바꾸었다.)

⑤ R/up eng.

(엔진을 정상 항행 상태로 하였다.)

⑥ Started F.W.G.

(조수기를 가동시켰다.)

3. At Sea(항해 중)

① Watched system oil pressure for M/E carefully.
(주기관의 시스템유 압력을 주의 깊게 관찰하였다.)

② Checked temperature & pressure of each part for M/E frequently.
(주기관 각부의 온도와 압력을 자주 점검하였다.)

③ Ascertained all b'rgs at normal temperature.
(모든 베어링의 온도가 정상인지를 확인하였다.)

④ Started No. 1 D.O purifier at 1500.
(오후 3시부터 No.1 경유 청정기를 가동했다.)

⑤ Adjusted consumption of cylinder oil to 100 *l* /day.
(실린더유의 소비량을 1일당 100 *l* 로 조정했다.)

⑥ Altered handle position of M/E from 60 to 40 at 0900.
(오전 9시에 M/E 핸들 위치를 60에서 40으로 변경했다.)

⑦ Soot Blower on for donkey boiler at 2000.
(오후 8시에 보조보일러의 그을음을 내뿜었다.)

⑧ Pumped out engine room bilge.
(기관실의 빌지를 배출했다.)

⑨ Refilled the cylinder lubricators with oil when the oil level had fallen by 50%.
(윤활유 레벨이 50%까지 내려갔으므로 실린더 주유기를 채웠다.)

4. Preparation of Arrival(입항 준비)

① Changed bunker 'C' to 'A'.
(벙커를 'C'에서 'A'로 바꾸었다.)

② Started No. 2 G/E and set Nos 1 & 2 generators in parallel running.
(2번 발전기를 가동시켜 1호기와 병렬 운전하였다.)

③ Stopped F.W.G.

(조수기를 정지시켰다.)

④ Arrived & anchored at quarantine station.
(검역 묘박지에 도착하여 투묘했다.)

⑤ Started to fumigate from stem to stern at 1000.
(10시부터 선내 소독을 시작했다.)

⑥ The ship docked in No.2 dock of K shipyard at 0930.
(9시 30분에 K 조선소 2호 도크에 입거했다.)

⑦ Commenced to pump out water of the dock at 1230.
(12시 30분에 도크의 물을 퍼내기 시작했다.)

⑧ Commenced to let water into the dry dock at noon.
(정오에 드라이 도크 속으로 물을 넣기 시작했다.)

5. In Port(정박 중)

① F.W.E.
(엔진을 휴지시켰다.)

② Opened test cocks of all cylinders for M/E and carried out air running.
(주기관의 모든 실린더의 테스트 콕을 열고 에어 러닝을 실시하였다.)

③ Stopped all pumps related to M/E.
(주기관과 관련된 모든 펌프를 정지시켰다.)

④ Started No. 2 G/E & stopped No.1 G/E at 1000.
(10시에 No. 2 발전기를 시동하고 No. 1 발전기를 정지시켰다.)

6. Maintenance(정비)

(1) Main Engine(주기관)

① Inspected all holding down bolts of M/E and found them in good condition.
(주기관의 지지 볼트 전부를 점검해 본 결과, 상태가 양호함을 확인하였다.)

② Overhauled No. 1 cyl. cover and inspected cooling surface.
(1번 실린더 커버를 분해하여 냉각 부위를 검사하였다.)

③ Renewed exhaust valve of No. 3 cyl.
(3번 실린더의 배기 밸브를 새 것으로 교체하였다.)

④ Replaced exhaust valve of No. 4 cyl.
(4번 실린더의 배기 밸브를 교체하였다.)

⑤ Disassembled and lapping 3 sets of exhaust valve and reassembled them.
(배기 밸브 3세트를 분해하여 연마하고 다시 조립하였다.)

⑥ Overhauled 3 sets of exhaust valve.
(배기 밸브 3세트를 분해 검사하였다.)

⑦ Took out No. 2 cyl. liner and repaired cracked parts.
(2번 실린더 라이너를 빼내어 균열 부위를 수리하였다.)

⑧ Measured amount of wear of cyl. liner.
(실린더 라이너의 마멸량을 계측하였다.)

⑨ Measured crank deflection.
(크랭크 디플렉션(개폐량)을 계측하였다.)

⑩ Piston rings, rings grooves, & cylinder top clearance to be measured.
(피스톤 링, 링 홈 및 실린더 톱의 간극을 계측했다.)

(2) Generator Engine(발전기 엔진)

① Overhauled emergency stop apparatus of No. 1 G/E and found it in good condition.
(1번 발전기 엔진의 비상 정지 장치를 분해 검사한 결과, 이상이 없음을 확인하였다.)

② Replaced F.O valves of all cylinders and 2 sets of exhaust valve of Nos. 1 & 4 cylinders.
(모든 실린더의 연료 밸브와 1번 및 4번 실린더의 배기 밸브 2세트를 교환하였다.)

③ Adjusted injection timing of No. 3 G/E.
(3번 발전기 엔진의 연료 분사시기를 조정하였다.)

④ Readjusted tappet clearance of No. 2 G/E.

(2번 발전기 엔진의 밸브 간극을 재조정하였다.)

(3) Boiler(보일러)

① Set fires on aux. boiler at 1900.

(19시에 보조 보일러를 점화시켰다.)

② Put out fires on aux. boiler.

(보조 보일러를 소화시켰다.)

③ Overhauled and lapped 3 pieces of steam stop valve for soot blower.

(그을음 제거기용 증기 정지 밸브 3개를 분해 검사하고 연마하였다.)

④ Replaced burner set with new one.

(버너 세트를 새 것으로 교체하였다.)

⑤ Carried out bottom and surface blow.

(수저(水低)·수면 방출을 실시하였다.)

⑥ Cleaned burner tip and oil strainer.

(버너 팁과 연료 여과기를 청소하였다.)

⑦ Boiler to be cleaned internally & externally for survey.

(검사받기 위하여 보일러 내부와 외부를 정비했다.)

⑧ Broken fire bricks in furnace mouth to be renewed.

(화로 아궁이의 파손된 내화 벽돌을 신품과 교환했다.)

⑨ Boiler water spaces to be chemically cleaned.

(보일러 수부(水部)를 화학 세척했다.)

(4) Purifier(청정기)

① Overhauled and cleaned separation disc of L.O purifier.

(윤활유 청정기의 분리판을 분해·청소하였다.)

② Changed gravity disc of No. 1 F.O purifier to dia. 100mm.

(1번 연료유 청정기의 비중판을 지름 100mm로 교환했다.)

(5) 냉동기(refrigerator)

① Opened cyl. cover of No. 1 comp'r for refrigerating machine and renewed all suction & delivery valves.
(냉동기 1번 압축기의 실린더 커버를 개방하여 모든 흡입 및 토출 밸브를 새 것으로 교환하였다.)

② Drew out all pistons of No. 1 air comp'r and renewed all piston rings and oil scraper rings.
(1번 공기 압축기의 모든 피스톤을 빼내어 피스톤링과 오일링을 전부 새 것으로 교환하였다.)

③ Charged 10kgf of Freon R-12 into refrigerating system.
(냉동 계통에 프레온 R-12를 10kgf 보충하였다.)

④ Made close examination for leakage of gas on all gas system for refrigerator.
(냉동기의 모든 가스 계통에 대하여 가스의 누설 여부를 자세히 조사했다.)

(6) Electric equipment(전기 장비)

① Measured insulation resistance of No. 1 cargo winch motor's winding.
(1번 하역 윈치 모터의 권선에 대하여 절연 저항을 측정하였다.)

② Tested all electrical circuits of generator, motors & lighting systems with megger.
(발전기, 모터 및 조명 장치의 모든 전기 회로를 메거로 점검하였다.)

③ Renewed 10 pieces of fluorescent lamp and 7 pieces of lamp in E/R.
(기관실 내 형광등 10개와 백열등 7개를 새 것으로 교환하였다.)

(7) Others(기타)

① Cleaned air filter of air conditioner.
(공기 조화 장치의 공기 여과기를 청소하였다.)

② Cleaned suction & delivery strainers of L.O pump.
(윤활유 펌프의 흡입 및 토출측 여과기를 청소하였다.)

③ Opened both end covers of L.O cooler for M/E and cleaned S.W. side.

(주기관용 윤활유 냉각기의 양쪽 커버를 열어서 해수측을 청소하였다.)

④ Renewed 4 pieces of zinc plates for auxiliary condenser.

(보조 응축기용 아연판 4개를 새 것으로 교환하였다.)

⑤ Lapped steam inlet valve for F.O heater of M/E.

(주기관 연료유 가열기의 증기 입구 밸브를 연마하였다.)

⑥ Washed and painted upper casing part in E/R.

(기관실의 상부 덮개를 물로 씻어 내고 페인트칠을 하였다.)

⑦ Cleaned and painted inside of purified room.

(청정기실 내부를 청소하고 페인트칠을 하였다.)

⑧ Replenished 500 liters of system oil into M/E sump tank.

(주기관용 섬프 탱크에 시스템유 500리터를 보충하였다.)

⑨ Started receiving F.O from 'K' Oil Co.

('K' 정유 회사로부터 연료유 수급을 시작하였다.)

⑩ Loaded 200 tons of F.O.

(연료유 200톤을 적재하였다.)

⑪ Took in 500 tons of H.F.O & 80 tons of M.D.O from 'G' Oil Co.

('G' 정유사로부터 H.F.O 500톤과 M.D.O 80톤을 수급하였다.)

⑫ Oiler, Mr. Lim joined her today.

(조기수 임씨가 오늘 승선하였다.)

⑬ Mr. Lee, 1/E left her today.

(1급 기사 이씨가 오늘 하선하였다.)

⑭ Being national holiday, no work.

(국경일이기 때문에 일과가 없다.)

⑮ Being New Year's Day, no work done.

(설날이므로 작업이 없었다.)

⑯ Advanced one hour for S.M.T.

(선내 평균시에 맞추기 위해 1시간 앞당겼다.)

⑰ Retarded 30 minutes for L.S.T.

(지방 표준시에 맞추기 위해 30분 늦추었다.)

⑱ Put clock 30 minutes ahead.

(시계를 30분 앞당겼다.)

⑲ Put clock 30 minutes back.

(시계를 30분 늦추었다.)

기본어휘

어휘	뜻	어휘	뜻
commence	시작하다	warming-up	워밍업[12)]
engage	맞물리게 하다	disengage	연결(접속)을 풀다
G/E[13)]	발전기	parallel running	병렬 운전
confirm	확인하다	S/B eng.	stand-by engine
S/B	stand-by, 준비	change	변화시키다
R/up eng.[14)]	정상 항행	F.W.G[15)]	조수기
watch	관찰하다	Jacket C.F.W[16)]	자켓 냉각 청수
adjust	조절하다	ascertain	확인하다
b' rg(bearing)	베어링	F.W.E	Finished with Engine
carry out	실행(수행)하다	air running[17)]	에어 러닝
inspect	점검(검사)하다	in good condition	상태가 양호한
surface	표면	renew	새 것과 교환하다
replace	교환하다	disassemble	분해하다
reassemble	다시 조립하다	overhaul	분해 검사하다
take out	빼내다	measure	측정하다
amount of wear	마모량, 마멸량	deflection	휨, 개폐량
injection timing	연료 분사 시기	readjust	재조정하다
clearance	여유, 틈, 간극	aux.(auxiliary)	보조의
set fires on	점화하다	put out	소화하다, 끄다
soot blower	그을음 제거기	carry out	실시하다

어휘	뜻	어휘	뜻
bottom blow[18)]	수저 방출	surface blow[19)]	수면 방출
dia.(diameter)	직경	comp'r (compressor)	압축기
refrigerator machine	냉동기	suction	흡입
delivery	방출, 토출	draw out	빼내다
kgf(kg force)	kg중	filter	여과기
air conditioner	공기 조화 장치	zinc plate	아연판
lap	갈다	insulation	절연(체)
resistance	저항	electrical circuit	전기 회로
megger	메거(절연 저항 시험기)	fluorescent lamp	형광등
replenish	보충(보급)하다	Co.(company)	회사
load	적재하다	take in	받아들이다
advance	앞당기다	S.M.T[20)]	선내평균시
retard	늦추다	L.S.T[21)]	지방표준시

12) 기관을 운전하기 전에 자켓 냉각수로 실린더의 온도를 높이는 일
13) Generator Engine
14) S/B eng.이 끝나고 일정한 속도로 항진하는 상태
15) Fresh Water Generator
16) Jacket Cooling Fresh Water
17) 압축 공기로 엔진을 잠시 동안 운전하는 것
18) 보일러 내부 바닥의 찌꺼기를 선외로 방출하는 것
19) 보일러 수상부의 부유 물질을 선회로 방출하는 것
20) Ship's Mean Time
21) Local Standard Time

Exercise

Ⅰ. 다음의 용어에 해당하는 우리말을 보기에서 고르세요.

(1) engine log book
(2) Pacific Ocean
(3) auxiliary machinery
(4) bearing
(5) delivery valve
(6) pilot
(7) M/E
(8) turning gear
(9) G/E
(10) S/B
(11) fuel valve
(12) current
(13) insulation resistance
(14) load
(15) R/up
(16) deflection
(17) lapping
(18) viscosity
(19) outlet temperature
(20) suction strainer
(21) steering gear
(22) compressor
(23) zinc plate
(24) purifier

보기

① 개폐량	② 흡입축 여과기	③ 연마	④ 점도	⑤ 도선사
⑥ 주기관	⑦ 토출 밸브	⑧ 터닝 기어	⑨ 준비	⑩ 부하
⑪ 아연판	⑫ 정상항행상태	⑬ 출구 온도	⑭ 태평양	⑮ 청정기
⑯ 보조기계	⑰ 조타 장치	⑱ 연료 밸브	⑲ 베어링	⑳ 압축기
㉑ 절연 저항	㉒ 발전기 엔진	㉓ 기관 일지	㉔ 전류	

Ⅱ. 다음 용어를 우리말로 바꾸어 보세요.

(1) warming-up
(2) parallel running
(3) cargo winch
(4) carry out
(5) exhaust valve
(6) cylinder liner
(7) gravity disc
(8) overhaul
(9) surface blow

Ⅲ. **빈 칸에 들어갈 알맞은 말을 본문에서 찾아 써 넣으세요.**

(1) Warmed up M/E (   ) 2 hours.

(2) (   ) turning gear and turned M/E for 20 minutes.

(3) (   ) bunker ‘A’ to ‘C’.

(4) (   ) No. 2 cyl. liner and repaired cracked parts.

(5) (   ) tappet clearance of No. 2 G/E.

(6) Oiler, Mr. Kim joined (   ) today.

(7) (   ) burner set with new one.

(8) (   ) 500 liters of system oil into M/E sump tank.

(9) (   ) suction & delivery strainers of L.O pump.

(10) (   ) one hour for S.M.T.

03

Repairing Orders
수리 주문

3.1 Entries of Repair Specification (수리 명세서 기재 내용)

1. Items of Repair Specification(수리 명세서 기재 항목)

수리 명세서의 양식은 회사에 따라 조금씩 다르나 기재 내용은 대개 비슷하다. 입거 수리와 검사 또는 항해 중 기기에 손상이 발생했을 때 수리 명세서에 기재할 사항을 열거해 보면 다음과 같다.

① 기기명

② Maker / Type / Serial No.

③ 결함 발생 날짜, 장소, 원인 및 내용

④ 최종 수리 날짜 및 내용

⑤ 수리할 내용

⑥ 수리 희망 장소 및 일시

⑦ 자재 보유 현황

⑧ 점검 결과 및 본선 수리 불가 사유

⑨ 관련 도면

2. 수리 명세서의 문장 특성

수리와 입거 관계의 수리 명세서에 사용되는 문장은 주로 명령문을 사용하고, '동사 원형+목적어' 또는 '목적어 + to + be + 과거분사형'을 취한다.

➡ Ex) Renew No. 1 piston ring.
Overhaul main thrust bearing.
Sea water pump to be overhauled.
Main bearing to be overhauled.

3.2 Examples of Entries in Repair Specification (수리 명세서 작성 예문)

1. Main Engine(주기관)

① No. 1 crosshead b'rg and crank pin b'rg to be overhauled for survey and oil clearance to be adjusted.
(검사에 대비하여 1번 크로스헤드 베어링과 크랭크핀 베어링을 개방 점검하고, 유간극을 조정할 것.)

② One length of starting air pipe to be renewed.
(시동 공기 파이프 1개를 새 것으로 교환할 것.)

③ The following parts to be repaired;
(다음 부분을 수리할 것.)

(a) Face up high pressure pipes/fuel pumps.
(고압 파이프와 연료 펌프를 연마하여 맞출 것.)

(b) Face up fuel valve spindles/guide bushes.
(연료 밸브 스핀들과 가이드 부시를 연마하여 맞출 것.)

(c) Face up fuel valve atomizer.
(연료 밸브 분무기를 연마하여 맞출 것.)

④ No. 1 crosshead bearing & crank bearing to be overhauled for survey, & oil clearance to be adjusted.
(1번 크로스헤드 베어링과 크랭크 베어링은 검사를 위하여 개방 검사하고 기름 간극을 조정할 것.)

⑤ Steel foot liners to be fitted to adjusted piston clearance.
(피스톤 간극을 조정하기 위하여 강철제 풋 라이너를 달 것.)

⑥ Nos. 2, 4 & 6 main bearings to be taken out for survey & oil clearance to be adjusted.

(2번, 4번 및 6번 메인 베어링은 검사를 위하여 빼내고 기름 간극을 조정 할 것.)

⑦ No. 5 fuel pump of main engine to be overhauled & valve timing to be adjusted.

(주기관 5번 연료 펌프를 개방 검사하고 밸브 타이밍을 조정할 것.)

2. Propeller & Shafting(추진기와 축계)

① Propeller blades to be polished and surveyed.

(프로펠러 블레이드를 연마하고 검사할 것.)

② Clearance between stern tube and tail shaft to be measured.

(선미관과 선미축 사이의 간극을 계측할 것.)

③ Worn zinc plates to be renewed.

(마모된 아연판을 새 것으로 교환할 것.)

④ Nos. 2 & 3 intermediate shaft bearings to be opened for survey and refitted on completion.

(검사에 대비하여 2번 및 3번 중간축 베어링을 개방하고, 종료 후 복구할 것.)

⑤ Propeller and outside fastenings to be examined.

(프로펠러 및 바깥쪽 잠금 장치를 검사할 것.)

⑥ Main thrust bearings to be overhauled for survey and thrust pads clearance to be checked and adjusted.

(검사에 대비하여 주 추력 베어링을 개방 점검하고, 추력 패드 간극을 검 사하고 조정할 것.)

3. Generator Engine(발전기)

① No. 2 G/E to be overhauled and repaired as follows and to be surveyed.

(2호 발전기를 개방 점검하고, 다음 개소를 수리한 후 검사받을 것.)

(a) All pistons to be overhauled.

(모든 피스톤을 개방 검사할 것.)

(b) All inlet valves to be lapped.

(모든 흡기 밸브를 연마할 것.)

(c) No.1, 2 & 3 cylinder liners to be renewed & corroded parts of the cylinder jacket to be repaired.

(1번, 2번, 3번 실린더 라이너를 신품과 교환하고 실린더 재킷의 부식된 부분을 수리할 것.)

② Two complete sets of fuel valve to be supplied as spares.

(연료 밸브 완비품 2세트를 예비품으로 공급할 것.)

4. Boiler(보일러)

① Boiler to be cleaned internal and external sides for survey.

(검사에 대비하여 보일러 내·외부를 소제할 것.)

② Boiler water space to be chemically cleaned.

(보일러의 수실을 화공 약품으로 소제할 것.)

③ Deformation of furnace to be measured.

(화로의 변형을 계측할 것.)

④ Fire bricks in furnace of boiler to be renewed as necessary.

(보일러 화로 내의 내화 벽돌을 필요에 따라 신품과 교환할 것.)

⑤ All valves of boiler mountings(12) to be overhauled & faced up.

(보일러 부속장치에 있는 모든 밸브(12개)를 분해하고 반듯하게 다듬을 것.)

5. Cooler, Condenser, etc.(냉각기, 응축기 등)

① Aux. condenser tube to be cleaned in cooling water side.

(보조 응축기 튜브의 냉각수측을 청소할 것.)

② Air and water sides of Nos. 1 & 2 air coolers to be cleaned and zinc plates to be renewed.

(1호 및 2호 공기 냉각기의 공기측과 수측을 청소하고 아연판을 새 것으로 교환할 것.)

6. Air Compressor(공기 압축기)

① Two sets of L.P. piston packing ring and two sets of H.P. piston packing ring to be renewed.

(저압 피스톤 패킹 링 2세트와 고압 피스톤 패킹 링 2세트를 새 것으로 교환할 것.)

② One sets of air compressor delivery valves to be fitted.

(공기 압축기의 토출 밸브 1세트를 설치할 것.)

7. Pump(펌프)

① L.O pump to be overhauled for survey.

(검사에 대비하여 윤활유 펌프를 개방 점검할 것.)

② Cooling S.W pump to be overhauled for survey.

(검사에 대비하여 냉각 해수 펌프를 개방 점검할 것.)

③ Pump to be reassembled and tested under working conditions.

(펌프를 다시 조립하여 작동하고 있는 상태에서 시험할 것.)

④ Impeller shaft of sanitary pump to be renewed & axial clearance of impeller to be adjusted.

(위생 펌프의 임펠러 축을 신품과 교환하고 임펠러의 축방향 간극을 조정할 것.)

8. Fresh Water Generator(조수기)

The following F.W generator valves to be overhauled.

(다음의 조수기 밸브를 개방 검사할 것.)

① feed valve(급수 밸브)

② ejector valve(이젝터 밸브)

③ flow meter valve(유량계 밸브)

9. Pipe & Valve(파이프와 밸브)

① Outlet valve of main cooling water pump to be repaired.
(주 냉각수 펌프 출구 밸브를 수리할 것.)

② Lagging on main steam pipes to be repaired.
(주 증기관의 단열재를 수리할 것.)

③ Cooling water inlet pipe to condenser to be repaired(attached to condenser).
(응축기 냉각수 입구 파이프(응축기에 부착된 것)를 수리할 것.)

④ Port side F.O filling pipe to be repaired and tested at 4kgf/cm2.
(좌현 연료유 급유관을 수리하고, 4kgf/㎠의 압력으로 시험할 것.)

⑤ All sea valves to be cleaned internally & coated with Apexior.
(모든 선외 밸브의 내부를 청소하고 어펙서를 도장할 것.)

⑥ All sea connections to be overhauled for survey.
(수검을 위하여 모든 해수도관을 개방 검사할 것.)

10. Electric equipment(전기 장비)

① Generators, electric motors & lighting circuit, etc. to be megger-tested & reports of readings before & after repair to be submitted to surveyor & our company's representative.
(발전기 · 전동기 및 전등회로 등을 메거 시험을 하고 수리 이전과 이후의 눈금의 기록을 검사관과 본사측 감독에게 제출 할 것.)

② One armature & three sets of field coils to be recoiled.
(전기자 1개와 계자 코일 3세트를 다시 감을 것.)

③ No.1 automatic circuit breaker to be repaired & adjusted.

(No.1 자동 회로 차단기를 수리하고 조정할 것.)

④ Two engine room ventilator blower motors to be overhauled on site, all parts to be cleaned, commutator to be polished, brushes to be beded in & adjusted.
(2대의 기관실 통풍기를 그 자리에서 개방 검사하고 모든 부품을 정비할것, 정류자를 반듯하게 놓고 조정할 것.)

⑤ Supplied shore electric power(AC 240V × 200A).
(육전을 공급받았다.(교류 240V × 200A))

11. Engine Room & Tank(기관실과 탱크)

① E/R and boiler room to be cleaned on completion of repairs.
(수리 완료 후 기관실 및 보일러실을 청소할 것.)

② Pneumatic tank depth gages for eleven fuel tanks to be overhauled, all gages assembly to be cleaned.
(11개의 연료유 탱크용 공기식 탱크 측심기를 개방 검사하고, 모든 계기 부품을 청소할 것.)

③ Engine room to be washed down & given a coat of paint. Paint ex ship.
(기관실을 청소하고 페인트를 1회 칠할 것, 페인트는 본선에서 지급함.)

④ Lubricating oil drain tank in cofferdam to be cleaned out & surveyed internally only.
(코퍼댐 내의 윤활유 드레인 탱크를 내부만 청소하고 검사할 것.)

⑤ Fuel oil tanks(not part of ship's structure) to be gas-freed, cleaned out for surveyor internal examination only.
(연료유 탱크(선체구조 이외의)를 검사관이 내부검사만 하도록 청소하고 가스를 제거할 것.)

⑥ Pneumatic tank depth gauges for ten fuel tanks to be overhauled, all gauge assembly to be cleaned.
(10개의 연료유 탱크용 공기식 탱크 측심계를 검사하고 모든 계기의 부품을 정비할 것.)

3.3 Examples of Entries in Accidents (각종 사고 예문)

1. Steering gear(조타기)

① Steering gear went wrong. Stopped engine & repaired it.
(조타기가 고장 났다. 기관을 정지시키고 수리했다.)

② Electric steering gear being disabled, changed to hand gear.
(전동 조타장치가 고장 나서 수동으로 전환했다.)

2. Propeller(추진기)

① Anchored temporarily on account of the accident to the propeller.
(추진기 사고로 인하여 가박(假泊)했다.)

② Had a diver remove the fouled rope & examined the propeller.
(잠수부가 얽힌 로프를 풀고 추진기를 시운전했다.)

③ Reported something unusual with her propeller to C/E.
(추진기의 이상을 기관장에게 보고했다.)

3. Fire(화재)

① Fire broke out in No.2 hold.
(2호창에서 화재가 발생했다.)

② Found fire in No.3 hold by the alarm of fire detector.
(화재탐지기의 경보로 3호창의 화재를 발견했다.)

③ Discovered smoke coming out of No.3 hold.
(3호창으로부터 연기가 나오는 것을 발견했다.)

④ The fire was due to falling down soots from the chimney of galley.
(화재는 취사실 굴뚝으로부터 떨어진 그을음이 원인이었다.)

4. Collision, Contact & Flooding(충돌 · 접촉 및 침수)

① M.S Ocean struck to our ship on the portside of abreast No.3 hold at about right angle with her stem.
(기선 오션호의 선수가 본선 3호창 좌현 정횡을 거의 직각으로 충돌했다.)

② The ballast pump threw out fresh water from FPT.
(밸러스트 펌프로 선수탱크의 청수를 퍼내었다.)

③ Started the bilge pumps to prevent foundering.
(침몰을 막기 위하여 빌지 펌프를 시동했다.)

④ Found leakage at the seams of starboard outside plates in No.2 hold.
(2호창 우현 외판 심으로부터의 누수를 발견했다.)

⑤ Inspected all tanks & bilge, and found leakage of S.W into APT.
(모든 탱크와 빌지를 점검하여 APT(선미탱크)으로의 해수 누입을 발견했다.

⑥ Commenced pumping out of water with donkey ballast pump & bilge pump.
(보조 밸러스트 펌프와 빌지 펌프로 배수하기 시작했다.)

기본어휘

어휘	뜻	어휘	뜻
b'rg(bearing)	베어링	survey	검사하다
surveyor	검사관	renew	새 것으로 교환하다
high pressure	고압	face up	연마하여 면을 맞추다
shafting	축계	blade	날개
polish	연마하다	between A and B	A와 B 사이
intermediate shaft	중간 축	completion	완성, 완료
outside	바깥쪽, 외면	fastening	죔, 죄는 용품
examine	검사(조사)하다	internal	내부의
external	외부의	chemically	화학적으로
deformation	변형	furnace	노
brick	벽돌	if necessary	필요하면
Aux.(auxiliary)	보조의	condenser	응축기
cooler	냉각기	L.P(Low Pressure)	저압
H.P(High Pressure)	고압	pp(pump)	펌프
lagging[22]	단열재	attach	붙이다
port	좌현	stb'd(starboard)	우현
filling pipe	급유관	pneumatic	공기의
depth ga(u)ge	측심기[23]		

22) 보온 및 안전을 위해 고온 파이프 또는 기기 등에 덮어 씌우는 단열재
23) 깊이를 재는 게이지

Exercise

Ⅰ. 다음의 용어에 해당하는 우리말을 보기에서 고르세요.

(1) crank pin bearing
(2) valve spindle
(3) zinc plate
(4) intermediate shaft
(5) condenser

보기

① 응축기 ② 아연판 ③ 밸브 스핀들
④ 중간축 ⑤ 크랭크핀 베어링

Ⅱ. 다음 용어를 우리말로 바꾸어 보세요.

(1) high pressure
(2) stern tube
(3) overhaul
(4) thrust bearing
(5) internal side
(6) fire brick
(7) air cooler
(8) flowmeter
(9) feed valve
(10) cooling water

Ⅲ. 빈 칸에 들어갈 알맞은 말을 본문에서 찾아 써 넣으세요.

(1) One length of starting air pipe to be (   ).
(2) (   ) high pressure pipes/fuel pumps.
(3) Propeller and outside (   ) to be examined.
(4) No. 2 G/E to be (   ) and repaired as follows and to be surveyed.
(5) Boiler water spaces to be chemically (   ).
(6) Deformation of furnace to be (   ).
(7) Lagging on main steam pipes to be (   ).
(8) Port side F.O filling pipe to be repaired and (   ) at 4kgf/cm^2.
(9) Pump to be (   ) and tested under working conditions.
(10) E/R and boiler room to be cleaned on (   ) of repairs.

04

Instruction Book

취급 설명서

4.1 Characteristics of Sentence

4.2 Examples of Engine Instruction Book

4.1 Characteristics of Sentence (문장의 특징)

1. 주부 및 술부가 긴 문장이 많다.

The difference between the pitch of the propeller and the actual advance of the ship in one revolution is called slip.

(프로펠러가 한 바퀴 회전하였을 때 선박이 실제로 전진한 거리와 프로펠러 피치와의 차이를 슬립이라 한다.)

2. 전문 용어는 두 단어 이상으로 구성된 합성어가 많다.

The main purposes of lubricating oil system aboard diesel engine are as follows.

(선박용 디젤 엔진에 있어서 윤활 시스템의 주요 목적은 다음과 같다.)

① To reduce friction and wear in the engine.
(엔진의 마찰과 마멸 감소)

② To act as a sealant between the rings and liners.
(링과 라이너 사이의 기밀 작용)

③ To cool the internal parts of the engine.
(엔진의 내부 냉각)

3. 수동태 형식의 문장이 많다.

The exhaust valve is opened hydraulically and closed by means of air pressure.

(배기 밸브는 유압으로 열리고 공기압으로 닫힌다.)

4. 관계 대명사 및 관계 부사가 문장 중에 많이 사용된다.

The function of bevel gears is to transmit motion between shafts the center lines of which intersect.

(베벨 기어의 기능은 중심선이 교차하는 축 사이의 운동을 전달하는 것이다.)

기본어휘

어휘	뜻	어휘	뜻
difference	다름, 차이	pitch[24]	피치
advance	전진하다	revolution	회전
internal combustion engine	내연기관	external combustion engine	외연기관
combustion	연소	intersect	교차하다

24) 프로펠러나 나사가 한 바퀴 회전했을 때 전진한 거리

4.2 Examples of Engine Instruction Book (기관 취급 설명서의 실례)

1. Diesel Engine(디젤 엔진)

There are two classes of diesel engines, that is, two stroke and four stroke. The two stroke engines are the type to complete cycle of operation in every two strokes of a piston. This type of engine requires compressed air for starting. In the four stroke engine, the first down stroke of the piston draws in air, which is compressed on the upstroke. At the top of the stroke a jet of oil is sprayed in through an injector. The oil is ignited and the rapid expansion of the gas created by the explosion forces the piston down in the working or firing stroke. The next upstroke drives the waste gases out through the exhaust valve, and the cycle is complete.

디젤 엔진에는 두 종류, 즉 2행정과 4행정이 있다. 2행정 엔진은 피스톤의 매 2행정마다 작동 사이클을 완료하는 형식인데, 이 형식의 엔진은 기동을 위하여 압축 공기가 필요하다. 4행정 엔진에서는 피스톤의 첫 하강 행정시 공기를 흡입하여, 그 다음 상승 행정 때 압축시킨다. 이 압축 행정의 끝에 인젝터를 통하여 연료를 분사한다. 기름이 연소하여 이 폭발력에 의해 생성된 가스의 급격한 팽창이 작동 또는 연소 행정에서 피스톤을 하강시킨다. 그 다음의 상승 행정에서 배기 밸브를 통하여 폐가스를 몰아내고 사이클은 완료된다.

2. Boiler(보일러)

Two types of boilers are most common fire-tube boilers, containing long steel tubes through which the hot gases from the furnace pass and around which the water to be changed to steam circulates, and water-tube boilers, in which the conditions are reversed. Water is changed to steam in these continuous circuits and also is superheated in transit. This additional heating of the steam increases the efficiency of the power generating cycle. The safety valve is used to prevent explosions by releasing steam if the pressure becomes too great.

보일러는 두 가지 형식이 가장 보편적이다. 연관 보일러는 긴 강관을 통하여 노에서 발생한 고온의 연소 가스가 통과하며 이주위의 물이 증기로 변하여 순환한다. 그리고 수관 보일러는 연관 보일러와 상태가 반대이다. 물은 이러한 연속된 순환 과정에 의하여 증기가 되며 또한 통과 도중에 과열된다. 증기의 부가적인 가열은 동력 발생 사이클의 효율을 증가시킨다. 안전밸브는 보일러 압력이 과도해졌을 때 증기를 방출하여 보일러 파열을 방지한다.

3. Refrigeration(냉동)

The pressure of the refrigerant gas is increased in the compressor and it thereby becomes hot. This hot and high pressure gas flows into a condenser. Depending on the particular application, the refrigerant gas will be cooled either by air or water, because it is still at a high pressure. The liquid refrigerant is then distributed through various branch pipes until it reaches a control valve alongside an evaporator where the cooling is required. This control valve meters the flow of liquid refrigerant into the evaporator, which is at a lower pressure.

압축기에서 냉매 가스의 압력은 증가하며, 그로 인해 고온이 된다. 이 고온 고압의 냉매 가스는 응축기로 유입된다. 냉매 가스는 여전히 고압이므로 특별한 장치를 이용하여 냉매 가스를 공기나 물로써 냉각시킨다. 액화된 냉매는 여러 분지 파이프를 통해 분배되어 냉각을 필요로 하는 곳(냉장실, 냉동실 등)에 있는 증발기 옆에 부착된 조절 밸브에 이르게 된다. 이 조절 밸브에서 액체 냉매의 유량을 조절하여 증발기에 보내면 이곳에서 저압 상태로 된다.

4. Electrical Equipment(전기 설비)

The complete electrical plant on board ship is made up of power generation equipment, a distribution system and the many power utilizing devices. Electricity is used for the motor drive of many auxiliaries and also for deck machinery, lighting, ventilation and air conditioning equipment. A constant supply of electricity is essential for safe ship and machinery operation, and therefore standby or additional capacity is necessary together with emergency supply equipment.

선박에서의 전기 설비는 동력 발생 장비, 분배 시스템 및 다수의 동력 이용 기구들로 이루어져 있다. 전기는 다수의 보조 기기용 모터를 구동하는 데 사용되며, 갑판보기, 점등, 환기 및 공기조화기에도 사용된다. 전기를 지속적으로 공급하는 것은 선박의 안전과 기기의 운전에 필수 요건이므로, 비상 전력 공급 장치와 더불어 예비 혹은 부가적인 전력 공급 장치가 필요하다.

기본어휘

어휘	뜻	어휘	뜻
two stroke	2행정	complete	완료하다
expansion	팽창	explosion	폭발
exhaust	배출하다	fire-tube boiler	연관 보일러
circulate	순환하다	water-tube boiler	수관 보일러
transit	통과, 통행	prevent	~을 막다, 예방하다
condense	농축(응축)시키다	application	적용, 응용, 신청
distribute	분배(분류)하다	refrigerant	냉각제, 냉매
equipment	비품, 설비, 기구	auxiliary	보조의, 보조기계
deck	갑판	essential	필수의, 절대 필요한
machinery	기계(류)	emergency	긴급(비상) 사태

Exercise

Ⅰ. 다음의 용어를 우리말로 바꾸어 보세요.

(1) internal combustion engine

(2) compressed oil

(3) fire-tube boiler

(4) power generation equipment

(5) emergency supply equipment

Ⅱ. 빈 칸에 들어갈 알맞은 말을 본문에서 찾아 써 넣으세요.

(1) Large diesel engine is a type of (   ) engine.

(2) This type of engine requires a supply of (   ) for starting.

(3) The (   ) of a diesel engine is complicated than that of a gasoline engine.

(4) At the top of the stroke a jet of oil is (   ) in through an injector.

(5) Boiler is a device for generating (   ).

(6) A (   ) is driven by steam generated under pressure in a boiler.

(7) This control valve meters the flow of liquid refrigerant into the (   ), which is at a lower pressure.

(8) The complete electrical plant on board ship is made up of (   ), a distribution system and the many power utilizing devices.

05

Practical English Conversation for Marine Engineers

기관사 실무 회화

5.1 Bunkering

5.2 Supply of Lubricating Oil

5.3 Engine Repair Work

5.1 Bunkering (급유)

(B/S : Bunker Supplier, C/E : Chief Engineer)

B/S : Good afternoon.

(안녕하세요.)

I'm from K Oil Company in Hongkong.

(저는 홍콩에 있는 K 정유회사에서 왔습니다.

C/E : Good afternoon. I'm the chief engineer.

(안녕하세요. 저는 기관장입니다.)

B/S : You ordered 1,000 tons of H.F.O and 200 tons of M.D.O through your agent, didn't you?

(대리점을 통하여 중유 1,000톤과 선박 디젤유 200톤을 청구하셨지요?)

C/E : That's right.

(네, 맞습니다.)

By the way, can I see the oil specifications?

(그런데 기름 명세서를 봐도 되겠습니까?)

B/S : Here it is.

(여기 있습니다.)

Check it and sign here, please.

(확인하시고 여기에 서명하여 주십시오.)

C/E : How many tons do you pump per hour?

(시간당 몇 톤으로 펌프 이송하실 겁니까?)

B/S : 150 tons an hour.

(시간당 150톤입니다.)

C/E : How about 200 tons?

(시간당 200톤은 보낼 수 없습니까?)

B/S : No. It's impossible. We need around 8 hours. That's the pumping capacity limit.

Have you closed the scupper?

(아니, 그렇게는 못합니다. 약 8시간이 걸립니다만, 이것이 펌프용량의 한도입니다. 본선의 스커퍼를 막아 두었습니까?)

C/E : I'll check it again and let you know.

(다시 조사해 보고 알려 드리겠습니다.)

B/S : The barge alongside. Won't you come to the sounding?

(부선이 막 접현했구요. 사운딩에 입회하시죠.)

C/E : The second engineer will be there.

(2등 기관사가 갈 예정입니다.)

2nd engineer, what's the outcome of the sounding?

(2등 기관사, 기름량 조사 결과 기름량은 얼마인가요?)

2/E : 1,003 tons of H.F.O and 201 tons M.D.O, sir.

(기관장님, 중유 1,003톤과 선박 디젤유 201톤입니다.)

C/E : Well. Good!

(네, 좋습니다.)

Now, connect the hose for supply.

(자, 급유를 위해 호스를 연결하세요.)

Fasten the hose coupling tightly.

(호스 연결 부분을 단단하게 조이세요.)

Barge : O.K. Are you ready?

(좋아요, 준비가 되었습니까?)

2/E : Yes, all valves are opened and we are ready.

(네, 모든 밸브가 개방되었고 준비가 되었습니다.)

Now start with M.D.O first and start slowly please.

(자, 먼저 선박 디젤유를 이송하겠습니다. 천천히 시작해 주십시오.)

Barge : O.K. Now, here it goes!

(좋아요. 지금 이송합니다.)

2/E : Stop! Oil is spilling a little from the flange!

(중지! 플랜지에서 소량의 기름이 새고 있습니다.)

Fasten it again!

(다시 단단히 조이십시오!)

Barge : It's all right now.

(이제 되었습니다.)

2/E : Well, Let's start again.

(좋아요, 다시 시작합시다.)

Barge : Now, we have finished with the diesel oil.

(지금, 디젤유 이송을 끝냈습니다.)

Shall we start with the H.F.O?

(중유 이송을 시작할까요?)

2/E : Wait about 5 minutes. We have to change the tank.

(5분간만 기다리십시오. 기름 탱크를 바꿔야 합니다.)

Good. Ready. Start the H.F.O slowly!

(좋아요. 준비되었어요. 중유 이송을 천천히 시작하시오.)

Barge : O.K. Now comes the fuel oil!

(좋아요. 자, 연료유 이송을 시작합니다!)

2/E : All right.

(좋아요.)

Raise it to the maximum pumping rate.

(펌프 최대 이송율로 올리십시오.)

And near the end, slow it down again.

(그리고 이송이 끝날 쯤에서 다시 낮추시오.)

2/E : What's the balance?

(지금 얼마나 남았습니까?)

Barge : We have more than 500 tons.

(아직 500톤 이상이 남은 것 같습니다.)

2/E : What's the pumping rate an hour now?

(지금 펌프 이송율로 얼마입니까?)

Barge : 130 tons per hour.

(시간당 130톤 정도 됩니다.)

2/E : Raise it to 150 tons.

(150톤 정도로 올려주세요.)

Barge : All right.

(예, 알겠습니다.)

Barge : Now, we are coming to the end. Never shut the valve before air blowing.

(지금, 기름 이송이 끝나갑니다. 공기 주입 전에 밸브를 닫지 마십시오.)

2/E : I understand.

(알겠습니다.)

Barge : Now, it's finished. I will blow the air.

(자, 끝났습니다. 공기 주입을 하겠습니다.)

2/E : Good.

(좋아요.)

Barge : Finished. Can we separate the hose?

(끝났습니다. 호스를 분리해도 됩니까?)

2/E : Make sure not to spill the oil.

(기름을 흘리지 않도록 주의하십시오.)

B/S : Chief engineer, we have finished bunkering.

(기관장님, 급유를 끝냈습니다.)

C/E : Just a moment.

(잠시만요.)

My second engineer is measuring the amount.

(2등 기관사가 기름량을 측량하고 있습니다.)

2/E : Sir, I finished the measuring.

(기관장님 적재량을 계산했습니다.)

B/S : You checked the amount?

(적재량, 확인이 되었습니까?)

C/E : By our measure, it's correct for the M.D.O but 25 tons short for the H.F.O Isn't that strange?

(본선의 계산상으로는 M.D.O는 정확하지만 H.F.O는 25톤이 부족합니다.)

B/S : By no means. We confirmed if on shore before carrying it on the barge. I don't think there are any mistakes with our meter on shore. Have you ever corrected the trim and heeling of the ship?

(그럴리가 있습니까? 부선에 싣기 전에 육상의 계량기로 확인했습니다. 육상의 계량기에 착오가 있다고는 생각지 않습니다. 본선의 트림과 힐링을 수정했습니까?)

C/E : Of course we have. We also corrected the temperature coefficient, trim and heeling. When did you correct your meter on the shore?

(물론이죠, 온도의 환산과 트림, 힐링도 수정 다 됐습니다. 육상의 계량기는 언제 수정됐습니까?)

B/S : I can't tell you until I check back to my office, but we have bunkered many ships and nothing has happened like this.

(돌아가서 검사하기 전까지는 말할 수 없으나 지금까지 많은 선박을 보유해 왔고 이 같은 사고는 없었습니다.)

C/E : We have not had such a thing, either. And we checked it repeatedly many times.

(본선 역시 이 같은 사고는 없었고, 몇 번이나 다시 확인해 본 양입니다.)

B/S : If you don't sign the receipt, it may be trouble.

(영수증에 서명해주지 않으면 곤란합니다.)

C/E : Well, then, I'll sign after correcting the H.F.O 975 tons.

(음, 그러면 H.F.O 975톤으로 수정한 후 사인하겠습니다.)

B/S : Hardly! Didn't you check it already by sounding before bunkering?

(안됩니다. 급유전에 이미 측심으로 확인하지 않았나요?)

C/E : Anyway we have a shortage. Now, what about this: Leave the number 1,000 tons untouched, and add a remark reading, "By our measure, we received 975 tons, and it's necessary to check meter. So the exact amount shall be rechecked after inspection is made on the meter and tank on the barge."

(아무튼 부족합니다. 그럼 이렇게 하면 어떻습니까. 1,000톤의 숫자는 그대로 두고, "본선 계측으로는 975톤을 수령하였으나 계량기를 검사할 필요가 있으며, 정확한 양은 계량기와 부선측 탱크의 조사가 있은 후 확인되어야 함" 이라고 Remark를 집어넣으면 어떻습니까!)

B/S : Agreed. Can I have the sounding data of this ship?

(그렇게 하지요. 본선의 사운딩 Data를 얻을 수 있겠습니까?)

C/E : Sure. I will give this one in square. Now, it is signed.

(예, 당장 정서하여 드리지요. 자, 서명했습니다.)

B/S : Very well. I am sorry for the unsatisfactory results. Bon voyage.

(좋습니다. 이 결말에 대하여 만족을 드리지 못하여 죄송합니다. 안전 항해를 빕니다.)

C/E : Thank you.

(수고 많았습니다.)

5.2 Supply of Lubricating Oil (윤활유 보급)

C/E : Are you from the agency.

(대리점에서 오셨습니까?)

Agent : Yes, I am.

(그렇습니다.)

C/E : I called you because we need lubricating oil urgently.

(윤활유가 급해서 전화했습니다.)

Agent : We arranged for it. It was Gulf Oil?

(수배를 했습니다. Gulf oil 이었지요?)

C/E : Right.

(그렇습니다.)

Agent : At 1 p.m, tomorrow, the system oil is scheduled to arrive by truck and the hydraulic oil will be loaded into the drum on a truck. How is that?

(내일 오후 1시에 system oil 이 트럭으로 올 것이며, Hydraulic oil 은 트럭에 의해서 드럼으로 적재될 예정입니다. 본선의 형편으로는 어떻습니까?)

C/E : Considering our repairing schedule, the sooner, the better. We have a leak in high pressure pipe of the #3 winch.

(정비작업 일정으로 보아서는 빠를수록 좋습니다. 오늘 오후 3번 윈치 고압파이프의 누설이 있습니다.)

Agent : No oil spill on the surface?

(해면으로 oil spill 이 일어나지 않았습니까?)

C/E : No pol spill, but most of the hydraulic oil has leaked. Now we have no oil margin on board.

(oil spill 은 없었지만, 대부분의 Hydraulic oil이 누출되었습니다. 그래서 지금 본선에서는 여분의 oil dl 조금도 없습니다.)

Agent : You've got a stroke of bad luck amid the misfortune. The oil company schedules full today, and they can't change it. But I'll try once again.

(설상가상이군요. 저쪽 Oil company의 사정으로는 오늘은 예정이 꽉 차있답니다. 아마 변경할 수 없으리라 생각합니다만, 다시 한 번 알아보겠습니다.)

Agent : They can't change the schedule, but as promised, they will make it tomorrow morning.

(변경은 할 수 없으나, 처음 예정했던 대로 내일 오전은 틀림이 없다고 합니다.)

C/E : That's O.K. Maybe we'll not experience the same thing twice in a day. During night work, I have to tell the foreman to avoid excessive work, and post duty on the deck to pay attention to the oil pressure pipe.

(좋습니다. 설마 오늘 중으로 두 번 같은 일이 생기지는 않겠지요. 그리고 야간 작업 중에는 Foreman에게 이야기하여 무리하지 않도록 할 것이며, 유압파이프에 신경을 쓰도록 당직을 갑판에 세워야겠군.)

Agent : How old is this ship?

(본선의 선령은 얼마나 되었습니까?)

C/E : 15 years.

(15년입니다.)

Agent : Pretty old.

(꽤 오래 되었군요.)

C/E : But I became intimate with this ship. Starting from 1st engineer, I served as chief-engineer for the first time in my life and for the second time. I am now working for 5 years on the ship.

(그러나 나는 이 배에 정이 들었습니다. 1기사 때부터 시작해서 처음 이 배에 서 기관장을 했고, 그 후 다시 두 번 지금 5년째 이 배에 근무하고 있습니다.)

Agent : Every bit it has your marked on it.

(기관장님의 손때가 안 간 곳이 없겠군요.)

C/E : Not exactly. However, as much attention as I might pay, there are still some blanks.

(그렇지 않습니다. 아무리 신경을 써도 못 미치는 곳이 있습니다.)

Agent : They must be happy it has met such a great man like you.

(이 배는 행복하겠군요. 기관장님과 같은 분을 만나서 말입니다.)

C/E : Thank you. Well, we'll not have much trouble to load the drums as freeboard is not too high, anyway, I'll get notice to everybody to get well prepared. Hope you tell the company to prepare those hoses sufficiently and notify them of our filling pipe flange which is 4 inches wide.

(감사합니다. 아무튼 내일 아침에는 그렇게 늦지 않겠으니 드럼을 탑재하는 데 어려움이 없으리라 생각되지만, 미리 충분한 준비를 하도록 일러두겠습니다. 저쪽에도 충분한 Hose를 준비하도록 일러두고 본선 Filling 파이프 Flange가 4인치임을 알려주시기 바랍니다.)

Agent : All right. I'll do it.

(알겠습니다. 그렇게 말하겠습니다.)

5.3 Engine Repair Work (기관 수리 작업)

(D/E : Dock Engine, C/E : Chief Engineer)

D/E : Good morning, sir.
(안녕하세요, 기관장님)
I am Brown, dock engineer in charge of engineering.
(저는 기관 담당 조선소 기사 브라운입니다.)

C/E : Glad to see you. Have a seat, please.
(만나서 반갑습니다. 자리에 앉으시죠.)
We want you to draw out the No. 3 piston of the main engine.
(우리는 당신이 주기관 3번 피스톤을 발출하기를 원합니다.)
Have you had experience with the H engine?
(H 엔진에 대해 경험이 있으십니까?)

D/E : You can count on me.
(저를 믿으셔도 됩니다.)
Now, what is the situation?
(지금 상태가 어떻습니까?)

C/E : We are to head homeward to Korea with the engine as it is, but the blow-by is too excessive.
(지금 상태로 본국인 한국으로 항해할 예정이지만 가스 누출이 매우 심합니다.)

D/E : How much are the cylinder liners worn out?
(실린더 라이너가 얼마나 마모되었습니까?)

C/E : I think they must be worn out about two millimeters.
(2밀리미터 정도는 마모되었다고 생각합니다.)

D/E : Then, it must be caused mainly by worn-out piston ring.
(그러면 주된 원인은 피스톤 링의 마모이겠군요.)

C/E : I think so.
(저도 그렇게 생각합니다.)

D/E : Is there anything for me to be careful about while doing the job?

(일을 진행하는 데 주의해야 할 점이 있습니까?)

C/E : I want you to be careful about setting up the stuffing box of the piston rod.

(피스톤 로드의 스터핑 박스 조립시 주의해 주시기 바랍니다.)

If it is done incorrectly, it will cause leakage of oil.

(조립이 잘못되면 기름이 누설됩니다.)

D/E : I understand. Can you lend me some special tools?

(알겠습니다. 전용 공구를 빌려 주시겠습니까?)

C/E : Of course. The engineer in charge will help you, so you can ask him about the details.

(물론이죠. 담당 기관사가 도와 줄 것입니다. 그에게 자세한 것을 물어 보세요.)

D/E : Very good, sir. What time are you leaving?

(좋습니다. 기관장님, 언제 출항하십니까?)

C/E : At eight o'clock tomorrow morning. So you should finish it by tonight.

(내일 아침 8시입니다. 그래서 오늘 밤에 작업을 끝내야 합니다.)

D/E : I am sure. I can finish it by ten o'clock tonight.

(알겠습니다. 오늘 밤 10시까지 끝낼 수 있습니다.)

Dock : Is the cooling water drained out?

(냉각수를 배출하였습니까?)

Ship : Yes, It is. All is ready now.

(네, 지금 모든 준비가 다 되었습니다.)

Dock : Shall I measure the clearance between the new ring and the groove?

(새로운 링과 홈 사이의 간격을 계측할까요?)

Ship : Yes, please.

(네, 그렇게 해 주십시오.)

Dock : Have you finished checking the cylinder covers?

(실린더 커버 점검은 끝났습니까?)

Ship : Yes, I have.

(네, 끝났습니다.)

Replace the gaskets with the new ones.

(개스킷을 새 것으로 교환해 주십시오.)

Dock : O.K.

(좋습니다.)

Dock : Everything has been restored. Please feed the cooling water and the lubricating oil.

(모든 것이 복귀되었습니다. 냉각수와 윤활유를 공급하세요.)

Ship : I'm going to feed the jacket cooling water first.

(먼저 재킷 냉각수를 공급하겠습니다.)

Dock : O.K. Anywhere is leaking?

(됐습니다. 어디 새는 곳이 있습니까?)

Ship : Seems to be no leaking.

(새는 곳이 없는 것 같군요.)

I'm going to feed the piston cooling water and the lubricating oil.

(피스톤 냉각수와 윤활유를 공급하겠습니다.)

Is the safety valve open? I'll start turning.

(안전밸브는 열려 있습니까? 기관 회전을 시작하겠습니다.)

Dock : It's open. It's turning O.K.

(열려 있습니다. 기관 회전 상태가 좋습니다.)

Ship : The piston cooling water and the lubricating oil are O.K.

(피스톤 냉각수와 윤활유는 좋습니다.)

I'll feed the air next. Stay near the flange.

(다음으로 공기를 공급하겠습니다. 플랜지에 가까이 계십시오.)

It is apt to leak there.

(거기가 새기 쉽습니다.)

Dock : There seems to be no leaking.

(새는 곳이 없는 것 같습니다.)

C/E : Thank you.

(감사합니다.)

We have finished an hour earlier than I expected.

(제가 예상했던 것보다 한 시간 일찍 끝났습니다.)

D/E : Now, sign this certificate for the work completion.

(여기 작업 완료 확인서에 서명하여 주십시오.)

C/E : Is this all right the way I signed?

(서명한 것이 맞게 되었습니까?)

D/E : Yes, sir. That's enough. Thank you very much.

Exercise

Ⅰ. 다음 문장은 대화문이다. 빈 칸에 알맞은 말을 보기에서 골라 보자.

(1) I'm (　　　　　) K Oil Company in Hongkong.

(2) It is (　　　　　) to leak there.

(3) I'm (　　　　　) to feed the piston cooling water.

(4) Can I see the oil (　　　　　)?

(5) Won't you come to the (　　　　　)?

(6) (　　　　　) the hose coupling tightly.

(7) Stop! Oil is (　　　　　) a little from the flange!

(8) I can finish it (　　　　　) ten o'clock tonight.

(9) Wait about 5 minutes. We have to (　　　　　) the tank.

(10) Raise it to the (　　　　　) pumping rate.

(11) (　　　　　) you lend me some special tools?

(12) (　　　　　) much are the cylinder liners worn out?

(13) We want you to (　　　　　) out the No. 3 piston of the main engine.

(14) Glad to see you. Have a (　　　　　), please.

(15) We are to (　　　　　) homeward to Korea.

(16) I think they (　　　　　) be worn out about two millimeters.

(17) If it is done (　　　　　), it will cause leakage of oil.

(18) You (　　　　　) finish it by tonight.

(19) Is the cooling water (　　　　　) out?

(20) We have finish an hour (　　　　　) than I expected.

보기

① earlier	② how	③ sounding	④ should
⑤ introduce	⑥ by	⑦ from	⑧ specifications
⑨ drained	⑩ sure	⑪ apt	⑫ fasten
⑬ going	⑭ can	⑮ must	⑯ minute
⑰ draw	⑱ change	⑲ incorrectly	⑳ maximum
㉑ seat	㉒ spilling	㉓ head	㉔ good

II. 아래 각 문장의 빈 칸에 알맞은 말을 고라 보세요.

(1) A : Good afternoon. I'm (   ) K Oil Co. in Hongkong.

B : Good afternoon. I'm a chief engineer.

① about ② upon

③ with ④ from

(2) A : (   ), can I see the oil specifications?

B : Here it is, check it and sign here, please.

① By the way ② As a matter of fact

③ As it is ④ Once upon a time

(3) A : Won't you (   ) to the sounding?

B : The second engineer will be there.

① around ② come

③ throughout ④ beneath

(4) A : Chief engineer, we have finished bunkering.

B : (   ), my second engineer is measuring the amount.

① Excuse me ② All right

③ Just a moment ④ Hello

(5) A : Have you had experience with the Himsen engine?

B : You can (   ) on me.

① sure ② count

③ certain ④ smile

06

English Conversation for Daily Life
생활 영어

6.1 At the Airport and Flight

1. Vocabulary and Useful Expressions

purpose of trip : 여행목적

for sightseeing : 관광하기 위해

to visit my relatives : 친척을 방문하기 위해

on business : 사업상

baggage = luggage : 수화물

carry-on baggage : 휴대용 짐

birth date : 생년월일

nationality : 국적

passport : 여권

airline ticket : 항공권

vacant : 비어있음 / occupied : 사용 중

flight attendant : 승무원

seat belt = safety belt : 안전벨트

briefcase : 서류가방

have a reservation : 예약하다

airsickness : 비행기 멀미

airsick bag : 위생 봉투

motion sickness pill : 멀미약

boarding pass : 탑승권

I.D card : 신분증

blanket : 담요

pillow : 베게

magazine : 잡지

something to drink : 마실 것

one-way ticket : 편도 항공권 / round trip ticket : 왕복 항공권

window seat : 창가 쪽 좌석 / aisle seat : 복도 쪽 좌석

2. Patterns Drill

① What's the purpose of your trip? (여행 목적은 무엇인가요?)

I'm here **for sightseeing**. (관광을 위해서 이곳에 왔어요.)

I'm here to visit my relatives. (친척은 방문하기 위해서 이곳에 왔어요.)

I'm here on business. (사업상 이곳에 왔어요.)

② How many pieces of carry-on baggage do you have? (탑승용 짐은 몇 개인가요?)

I have only one. (단 한 개입니다.)

③ What's your nationality? (국적이 어딘가요?)

Korea. (한국입니다.)

④ Have **a nice trip**. (즐거운 여행되세요.)

Have a good time. (좋은 시간 되세요.)

⑤ How long are you going to stay in Canada?

(캐나다에 얼마나 머무를 예정인가요?)

I'm going to stay for 5 days. (5일 머무를 예정입니다.)

⑥ May I see your **boarding pass**? (탑승권을 보여주시겠습니까?)

airline ticket (항공권)

I.D card (신분증)

⑦ May I have a **blanket**? (담요가 있나요?)

pillow (베게)

magazine (잡지)

⑧ Would you like something to **drink**? (마실 것 좀 드릴까요?)

read

want

need

3. Dialogues

Dialogue 1

(O : Immigration Officer, P : Passenger)

O : May I see your passport?

(여권 좀 보여주시겠습니까?)

P : Here it is.

(여기 있습니다.)

O : What's the purpose of your visit?

(방문 목적이 무엇인가요?)

P : I'm here for sightseeing.

(관광차 왔습니다.)

O : How long are you going to stay in the States?

(미국에 얼마동안 체류할 예정인가요?)

P : (I'm staying for) about six days.

(약 6일 동안 체류할 예정입니다.)

O : Where will you be staying?

(어디에서 머무십니까?)

P : I will be staying at the Hilton Hotel.

(힐튼 호텔이요.)

Dialogue 2

(S : Stewardess, P : Passenger)

S : Would you like something to drink?

(마실 것 좀 드릴까요?)

P : Yes, what kinds of beverage do you have?

(예, 어떤 종류의 음료수가 있나요?)

S : We have coffee, tea, orange juice, coke and beer.

(커피, 홍차, 오렌지 주스, 콜라, 그리고 맥주가 있습니다.)

P : I'd like coffee, please.

(커피 주세요.)

S : How would you like your coffee? Cream or sugar?

(어떻게 드릴까요? 크림이나 설탕은요?)

P : Both, please.

(둘 다 주세요.)

S : Here they are.

(여기 있습니다.)

P : Thank you.

(감사합니다.)

4. Exercise

purpose of trip

for sightseeing

to visit my relatives

on business

baggage = luggage

carry-on baggage

birth date

nationality

passport

airline ticket

one-way ticket / round trip ticket

window seat / aisle seat

restroom, toilet

vacant / occupied

flight attendant

seat belt =safety belt

briefcase

have a reservation

airsickness

airsick bag

motion sickness pill

boarding pass

I.D card

blanket

pillow

magazine

something to drink

6.2 Getting Information

1. Vocabulary and Useful Expressions

sightseeing guidebook : 여행 안내 책자

timetable : 시간표

public transportation : 대중교통

taxi stand : 택시 타는 곳 / bus stop : 버스정류장

subway station : 지하철 역 / train station : 기차 역

change subway lines : 지하철을 갈아타다

get to ~ : ~에 도착하다

take a taxi(= catch a taxi) : 택시 타다

rent a car : 차를 렌트하다

traffic information : 교통 정보

How much ~ ? : ~이 얼마죠?

taxi fare : 택시 요금 / basic fare : 기본요금

What kinds of ~ ? : 어떤 종류의 ~

express bus terminal : 고속버스 터미널

international airport : 국제공항

How do I get to ~ ? : ~에 어떻게 가야하나요?

Can you tell me the way to ~ ? : ~가는 방법 알려주세요?

Is there a ~ near here ? : 이 근처에 ~ 이 있나요?

Where is the ~ ? : ~ 이 어디에 있나요?

go straight : 쭉 가세요.

turn left : 왼쪽으로 도세요.

next to~ : ~옆에

across from ~ : ~건너편에

on your right : 오른쪽에

in the middle of the ~ : ~의 가운데에

Where can I catch a taxi?

(어디서 택시를 탈 수 있나요?)

How long does it take to go downtown?

(시내까지 시간이 얼마나 걸리나요?)

I'm looking for a hotel around here.

(이 근처에 호텔을 찾고 있어요.)

Can I have a bus time table, please?

(버스 시간표 좀 얻을 수 있나요?)

Please tell us about the attractive places to visit in this area.

(이 지역의 명소 좀 알려주세요.)

Where is the nearest bus stop?

(가장 가까운 버스 정류장이 어딘가요?)

2. Patterns Drill

① Where can I take a **taxi**? (어디에서 택시를 탈 수 있나요?)

subway

express bus

train

② Could you tell me how to get to **the airport**? (공항 가는 방법 알려주세요?)

the express bus terminal (고속버스 터미널)

city hall (시청)

post office (우체국)

③ Is there a sightseeing guidebook for this area? (이 지역 안내책자가 있나요?)

Is there an art museum near here?

Is there a traditional market around here ?

④ Go straight this street about 200 meters and turn left.

(이 길을 따라 약 200미터 가서 좌회전 하세요.)

You can find it on your right. (오른쪽에 있어요.)

It's on the right.

next to the bakery

across from the bus stop.

I'm new (=a stranger) here.

(저도 이곳이 처음입니다.)

⑤ How long does it take to go there?

(그곳까지 얼마나 시간이 걸리나요?)

It takes about 30 minutes to go there.

3. Dialogues

Dialogue 1

A : Excuse me, officer. Where should I get a cab?

(실례합니다. 어디에서 택시를 탈 수 있나요?)

B : You may take a taxi at the front of a hotel.

(호텔 앞에서 탈 수 있어요.)

A : How much is the taxi fare?

(택시요금은 얼마인가요?)

B : The basic fare is 2,800 won.

(기본요금이 2,800원입니다.)

Dialogue 2

A : Excuse me. Where is the express bus terminal?

(실례합니다. 고속버스터미널이 어디에 있나요?)

B : It's located at Banpo dong in downtown.

(반포동에 있어요.)

A : How can I get there?

(어떻게 가야하나요?)

B : You'd better take a subway around here.

(이 근처에서 지하철을 타는 것이 좋겠습니다.)

Dialogue 3

(T : Tourist, P : Passerby)

T : Excuse me.

(실례합니다.)

P : Can I help you?

(도와 드릴까요?)

T : Yes. What's the best way to get to the Empire State Building?

(예. Empire State 빌딩에 갈 수 있는 가장 좋은 방법이 무엇인가요?)

P : It's a lot more convenient to take the bus.

(버스를 타시는 것이 훨씬 편해요.)

T : Can I get a bus around here?

(이 근처에서 버스를 탈 수 있나요?)

P : Yes, you can.

(예. 탈 수 있어요.)

T : How long does it take to get there by bus?

(시간이 얼마나 걸리나요?)

P : It takes about 20 minutes.

(대략 20분 정도 소요됩니다.)

T : Where is the nearest bus stop?

(가장 가까운 버스 정류장이 어딘가요?)

P : It's at the end of this block.

(이 블록 끝에 있습니다.)

T : Thank you very much.

(대단히 감사합니다.)

4. Exercise

sightseeing guidebook

timetable

public transportation

taxi stand / bus stop

subway station / train station

change subway lines

get to ~

take a taxi(= catch a taxi)

rent a car

traffic information

How much ~ ?

taxi fare / basic fare

What kinds of ~ ?

express bus terminal

international airport

How do I get to ~ ?

Can you tell me the way to ~ ?

Is there a ~ near here ?

Where is the ~ ?

go straight

turn left

next to

across from ~

on your right

in the middle of the ~

Where can I catch a taxi?

How long does it take to go downtown?

I'm looking for a hotel around here.

Can I have a bus time table, please?

Please tell us about the attractive places to visit in this area.

Where is the nearest bus stop?

6.3 Shopping

1. Vocabulary and useful expressions

How many + 복수명사 are + 주어 ~ ?
do + 주어 + 동사원형~ ?
How much (셀 수 없는 명사) ~ ?

How many children do you have? (아이들이 몇 명인가요?)
How many cups of coffee do you drink a day? (하루에 커피는 몇 잔 마시나요?)
How much water do you drink a day? (하루에 물은 얼마나 마시나요?)
How much is it ? (얼마죠?)
How much does it cost?

try on : 입어보다
fitting room : 탈의실
look for ~ : ~을 찾다
look good on ~ : ~에게 잘 어울리다
This shirt looks good on you.
How does it fit? 어떤가요?
It's too tight. 너무 작아요.
too loose : 너무 큰
look around : 둘러보다
small / medium / large size
copy = receipt : 영수증
Here is your copy. 영수증이 여기 있어요.

I'm just looking around. 그냥 둘러보고 있어요.

May I help you? 도와드릴까요?

May(=Can) I try it on? 입어 봐도 되나요?

souvenir : 기념품

recommend : 추천하다

I'll take it. 그것으로 사겠습니다.

over there : 저기에

How about ~? ~은 어떤가요?

2. Patterns Drill

① **May** I help you? 도와 드릴까요?

Can I help you ?

How can I help you?

② I'm looking for **some souvenirs**. 기념품을 찾고 있어요.

yellow shirt for my dad

convenience shoes for me

③ What size do you **wear**? 어떤 사이즈를 입나요?

need

like

want

④ May I try on **this dress**? 이 옷을 입어 봐도 되나요?

this ring

these shoes

the hat over there

⑤ Which **hat** do you recommend for me? 어떤 모자를 추천해주시겠습니까?

souvenir

shoes

dress

⑥ I'm just **looking around**. 그냥 둘러보고 있어요.

browsing shopping

window shopping

3. Dialogues

Dialogue 1

(S : Sales woman, G : Guest)

S : Are you looking for some souvenirs?

(기념품을 찾나요?)

G : Yes, I am. What kind of souvenirs do you have to take to Korea ?

(예. 한국에 갖고 갈 어떤 종류의 기념품이 있나요?)

S : For typical souvenirs, I recommend English tea, a rain coat, chinaware or something like that.

(일반적인 것으로, 영국 차, 비 옷, 도자기 같은 것을 추천해 드릴께요.)

G : Which of them do you recommend for me?

(그 중에서도 저를 위해 추천해주세요.)

S : English tea is the most popular as souvenirs.

(영국 차가 기념품으로는 가장 인기가 있어요.)

G : How much is this one?

(이것은 얼마입니까?)

S : 20 dollars.

(20달러입니다.)

G : Do you have a cheaper one?

(더 싼 것도 있나요?)

S : This is $10 per set.

(이것은 세트에 10달러입니다.)

G : Oh, really? This seems good for me. I'll take it.

(정말요? 이것이 저에게 적당한 것 같아요 이것으로 사겠습니다.)

Dialogue 2

(S : Sales women, G : Guest)

S : May I help you?

(도와 드릴까요?)

G : Yes. I'm looking for a jacket for myself.

(예. 제가 입을 재킷을 찾고 있어요.)

S : How about this one? I think it'll look good on you.

(이것은 어떤가요? 제 생각에는 손님에게 잘 어울릴 것 같은데.)

G : Yeah, I like it. Can I try it on?

(맘에 드네요. 입아 볼 수 있나요?)

S : Sure. The fitting room is over there.

(물론입니다. 탈의실은 저기에 있어요.)

Later (잠시 후)

S : How does it fit?

(잘 맞습니까?)

G : It's too tight. May I have a larger one?

(너무 꼭 끼입니다. 더 큰 것이 있나요?)

S : Here is half a size larger.

(여기 반 인치 더 큰 것이 있어요.)

G : Oh, this looks comfortable. I'll take it.

(이것이 편해 보여요. 이것으로 사겠습니다.)

4. Exercise

How many cups of coffee do you drink a day?

How much water do you drink a day?

How much is it ?

How much does it cost?

try on

fitting room

look for

look good on ~

This shirt looks good on you.

How does it fit?

It's too tight.

too loose

look around

small / medium / large size

copy = receipt

Here is your copy.

I'm just looking around.

May I help you?

May(=Can) I try it on?

souvenir

recommend

what kind of ~ ?

I'll take it.

How about this one ?

over there

Which hat do you recommend for me?

6.4 At the Restaurant

1. Vocabulary and Useful Expressions

How would you like your steak? (요리방법을 묻는 표현)

고기(meat) well done
medium
rare
overcooked (너무 익은)

계란(egg) soft-boiled
hard-boiled
sunny-side up
(tipped) over
scrambled

What would you like to have? (무엇을 드시겠습니까?)
I'll have T-bone steak.
I'd like T-bone steak.
T-bone steak is fine.

How many in your party? (몇 분이십니까?)
Do you have a table available? (빈자리가 있습니까?)
Can I have a menu? (메뉴판 좀 갖다 주실래요?)

today's special : 오늘의 특식
the most popular dish : 가장 인기 있는 요리
salty : 짠
bland : 싱거운

spicy : 매운

Continental breakfast : 빵과 우유 또는 주스가 제공되는 간단한 유럽식 조식

American breakfast : 미국식 조식

dessert : 후식

What kind of dessert do you have?

(어떤 종류의 후식을 드시겠습니까?)

2. Patterns Drill

① What did you have for **breakfast**?

lunch

dinner

brunch

② What kind of **food** would you like to have?

drink

wine

salad

③ How would you like to have your **steak**?

eggs

meat

④ I always enjoy(a cup of) **coffee** after work.

milk

wine

lemonade

3. Dialogues

Dialogue 1

(W : Waiter, G : Guest)

W : Good morning. May I take your order?

(안녕하십니까, 주문하시겠습니까?)

G : Yes, I'd like to have the Continental breakfast.

(예, 저는 유럽식 조식으로 식사하고 싶습니다.)

W : Would you like to have orange juice or milk?

(오렌지 주스로 드시겠습니까? 우유로 드시겠습니까?)

G : Orange juice, please.

(오렌지 주스로 주세요.)

W : Would you like coffee or tea?

(커피를 드시겠습니까? 홍차를 드시겠습니까?)

G : Coffee, please.

(커피로 주세요.)

W : Cream and sugar is on the table.

(크림과 설탕은 테이블 위에 있습니다.)

G : Thank you.

(감사합니다.)

Dialogue 2

W : Are you ready to order now?

(지금 주문하시겠습니까?)

G : Yes.

(예.)

W : What would you like to have?

(무엇을 드시겠습니까?)

G : I'll have a New York Steak.

(뉴욕 스테이크로 하겠습니다.)

W : How would you like it that done?

(어떻게 익혀 드릴까요?)

G : Well done, please.

(잘 익혀 주세요.)

W : Baked potatoes or rice?

(구운 감자로 하시겠습니까, 밥으로 하시겠습니까?)

G : Baked potato, please.

(구운 감자로 주세요.)

W : Soup or salad?

(수프를 드시겠습니까, 샐러드로 드시겠습니까?)

G : Salad, please.

(샐러드로 주세요.)

W : What kind of dressing would you like?

(드레싱은 뭐로 할까요?)

G : I'd like Thousand Island.

(사우전드 아일랜드로 주세요.)

W : Have you decided what kind of wine you'd like to have?

(무슨 포도주를 드실지 결정하셨어요?)

G : Thank you. I don't drink.

(고맙습니다만, 전 술을 안마십니다.)

Dialogue 3

W : How was your dinner?

(식사는 어떠셨습니까?)

G : It was very good.

(아주 좋았습니다.)

W : I'm glad. Would you like to have some coffee now?

(다행이군요. 지금 커피 드시겠습니까?)

G : Yes, please.

(예, 주세요.)

W : How about some dessert?

(후식 드시겠습니까?)

G : What would you suggest?

(뭐가 좋아요?)

W : How about a piece of cheesecake? It's excellent.

(치즈 케이크가 어떻습니까? 아주 좋습니다.)

G : That sounds good.

(그거 좋을 것 같군요.)

4. Exercise

How would you like your steak?

well done

medium

rare

overcooked

soft-boiled

hard-boiled

sunny-side up

(tipped) over

scrambled

What would you like to have?

I'll have T-bone steak.

I'd like T-bone steak.

T-bone steak is fine.

How many in your party?

Do you have a table available?

Can I have a menu?

today's special

the most popular dish

salty

bland

spicy

Continental breakfast

American breakfast

dessert

What kind of dessert do you have?

6.5 Telephone

1. Vocabulary and Useful Expressions

local call : 시내전화

long distance call : 시외전화

person to person call : 지명통화

credit card call : 신용카드 통화

station to station call : 전화번호 지정 통화

collect call : 수신인 부담 통화

overseas call : 국제 전화

Hello, is that the Accounts Department?
(여보세요, 거기 경리부인가요?)

May I speak to the Manager, please?
(매니저 좀 바꿔 주세요.)

Who's calling, please? = Who's speaking, please?
(전화거시는 분은 누구십니까?)

Hold the line, please.
(잠깐만 기다려주세요.)

I'm sorry I can't hear you. Can you speak a little louder?
(전화 상태가 안 좋은데 좀 더 큰 소리로 말씀해 주세요.)

I'll get him/her for you. = He/She will be right with you.
(곧 바꿔 드리겠습니다.)

I'm sorry you've got the wrong number.
(전화를 잘못 거신 것 같군요.)

I'm sorry he's busy.
(그 분은 지금 바쁘십니다(통화 중입니다).)

I'm sorry he's out.

(그 분은 지금 자리에 없습니다.)

Could you tell me when she'll be available?

(언제쯤 통화할 수 있겠습니까?)

Can / Could / May I take a message? = Would you like to leave a message?

(메모를 남기시겠어요?)

2. Patterns Drill

① **May I speak** to Mr. Brown?

Can I talk

I'd like

② May I **ask** who's **calling**?

know talking

speaking

③ **You have** the wrong number.

You've got

I'm afraid you've dialed

④ **Hold the line**, please.

Wait a minute

Just a moment

Hold on

⑤ **I'm sorry** he's **busy**.

I'm afraid tied up

occupied

on the line

on another line

⑥ I'm sorry he's **out**.

not in the office

just gone[stepped] out

not available now

on a business trip

out of town

3. Dialogues

Dialogue 1

A : Hello, Sales Department.

(여보세요, 영업부입니다.)

B : May I speak to Miss Jeong, please?

(미스 정 좀 바꿔 주시겠습니까?)

A : Miss Jeong? Oh, you have the wrong extension. She's on extension 268. Hold the line and I'll transfer you.

(미스 정이요? 아, 전화를 잘못 거셨군요. 미스 정은 내선 286번입니다. 잠깐만 기다리시면, 그 쪽으로 돌려 드릴게요.)

B : Thank you.

(감사합니다.)

Dialogue 2

A : Hello, Hanjin Travel Agency Co. Ltd. May I help you?

(네, 한진여행사입니다. 도와드릴까요?)

B : Is Miss Bae there?

(미스 배 있습니까?)

A : I'm sorry, I can't hear you. Can you speak a bit louder?

(죄송하지만, 소리가 잘 안 들리는데. 좀 큰 소리로 말씀해 주시겠습니까?)

B : Can I speak to Miss Bae, please?

(미스 배 좀 바꿔 주세요.)

A : We have a bad line. Could you hang up and call back?

(연결 상태가 좋지 않군요. 끊고 다시 걸어주시죠.)

B : Yes, I'll call back.

(예, 다시 전화 걸죠.)

Dialogue 3

A : Hello. Can I speak to Mr. Han, please?

(여보세요, 한 선생님 좀 바꿔 주세요.)

B : Who's calling, please?

(누구신가요?)

A : This is Mr. Seaton.

(미스터 시튼입니다.)

B : Hold the line, please. Mr. Seaton Hello, I'm afraid Mr. Han in conference at the moment.

(잠시만 기다려 주십시오, 시튼씨. 여보세요, 죄송합니다만 한 선생님은 지금 회의 중이신데요.)

A : Oh, I see. When will he be free?

(오, 알겠습니다. 언제쯤 통화할 수 있겠습니까?)

B : He should be free at 4 o'clock. Can I take a message?

(4시엔 시간이 나실 겁니다. 메모를 남기시겠어요?)

A : Yes. Would you ask him to call me back this afternoon?

(예, 오늘 오후에 제게 전화해달라고 전해주세요.)

B : Certainly. Can I have your name and number, please?

(그러죠. 성함과 연락처를 말씀해 주세요.)

A : This is Mr. Seaton and my number is 765-4434.

(이름은 시튼이고요, 번호는 765-4434입니다.)

B : Can you spell your name, please?

(성함의 철자를 말씀해 주세요.)

A : Yes. S-E-A-T-O-N.

(예, 에스-이-에이-티-오-앤입니다.)

B : Thank you, Mr. Seaton. I'll give Mr. Han the message.

(감사합니다, 시튼씨. 한 선생님께 전해드리죠.)

A : Thank you, bye.

(감사합니다. 그럼...)

B : Good bye.

(안녕히 계십시오.)

4. Exercise

local call

long distance call

person to person call

credit card call

station to station call

collect call

overseas call

Hello, is that the Accounts Department?

May I speak to the Manager, please?

Who's calling, please? = Who's speaking, please?

Hold the line, please.

I'm sorry I can't hear you. Can you speak a little louder?

I'll get him/her for you. = He/She will be right with you.

I'm sorry you've got the wrong number.

I'm sorry he's busy.

I'm sorry he's out.

Could you tell me when she'll be available?

Can/Could/May I take a message? = Would you like to leave a message?

6.6 Health

1. Vocabulary and Useful Expressions

headache : 두통　　cf) migraine 편두통

toothache : 치통

stomachache : 복통

backache : 요통

cough : 기침

cold : 감기

flue / influenza : 독감

a touch of cold / a slight cold : 감기 기운

fever : 열

fatigue : 피로

diarrhea : 설사

high blood pressure : 고혈압

diabetes : 당뇨

tuberculosis / consumption : 결핵

heart problem(disease) : 심장질환

eczema : 습진

athlete's foot : 무좀

sore : 아픈

painful : 아픈, 고통스러운

stiff : 뻣뻣한

runny : 콧물이 흐르는

pale : 창백한, 핼쑥한

throbbing : 욱신거리는, 지끈거리는

tired : 피곤한, 지친

break one's arm : ~의 팔이 부러지다

scratch one's finger : ~의 손가락을 할퀴다/긁히다

injure one's knee : ~의 무릎 부상을 당하다

hurt one's shoulder : ~의 어깨를 다치다

2. Patterns Drill

① You look **happy**.
tired
healthy
sleepy

② What's **wrong**?
the matter
up
happening

③ How **are you** feeling today?
are they
is he
is she

④ I have a **cold**.
bad cough
headache
sore throat

⑤ I've **broken my arm**.
scratched my finger
injured my knee
hurt my shoulder

3. Dialogues

Dialogue 1

A : You look like you're feeling under the weather.

(몸이 불편해 보이는데.)

B : I am. I feel simply awful.

(그래. 굉장히 안 좋아.)

A : So, where's the pain?

(그래, 어디가 아픈 거야?)

B : Well, my stomach is upset, and I have aches and pains all over.

(그저, 배가 아프고, 온 몸이 쑤셔.)

A : You should go home then.

(그럼 집에 가보는 게 좋겠군.)

B : Yeah, I'm taking the day off.

(그래, 오늘 하루 쉬려고 해.)

Dialogue 2

A : General Hospital. May I help you?

(종합병원입니다. 뭘 도와드릴까요?)

B : Yes. This is an emergency.

(예. 긴급한 일입니다.)

A : All right, sir. Please tell me the nature of the emergency, where you are, your name and telephone number.

(잘 알겠습니다. 저에게 상황을 설명해주시고, 당신이 있는 위치, 성함 그리고 전화번호를 알려주세요.)

B : I'm sorry. I didn't understand that. Could you repeat that, please?

(미안합니다. 무슨 말인지 잘 모르겠습니다. 다시 한 번 말씀해 주시겠습니까?)

A : Yes, sir. Tell me what happened.

(예. 무슨 일이 일어났는지 저에게 말씀해 주세요.)

B : My roommate is very sick. He is vomiting, has a fever, and aches all over.
(제 룸메이트가 굉장히 아픕니다. 토하고, 열도 나고, 그리고 안 쑤시는 데가 없답니다.)

A : I see. That could be salmonella.
(알았습니다. 살모넬라인 것 같습니다.)

B : Pardon? I didn't catch that last part.
(뭐라고요? 마지막 부분이 무슨 말인지 모르겠습니다.)

A : "Salmonella." That's food poisoning. Now please give me your address.
(살모넬라라고 했습니다. 말이지요. 자, 이제 당신의 주소를 말씀해 주세요.)

B : We've in Room 508 of the Essex Hotel at 2532 Simpson Avenue.
(심슨가 2532번지에 있는 에섹스 호텔 508호에 있습니다.)

A : All right. Have you friend lie down and an ambulance will be on the way immediately.
(알겠습니다. 당신의 친구를 눕혀 놓으시면 앰뷸런스가 곧 갈 것입니다.)

4. Exercise

headache cf) migraine

toothache

stomachache

backache

cough

cold

flue / influenza

a touch of cold / a slight cold

fever

fatigue

diarrhea

high blood pressure

diabetes

tuberculosis / consumption

heart problem(disease)

eczema

athlete's foot

sore

painful

stiff

runny

pale

throbbing

tired

break one's arm

scratch one's finger

injure one's knee

hurt one's shoulder

6.7 Sightseeing

1. Vocabulary and Useful Expressions

in bound : 외국인에 의한 국내여행

out bound : 내국인에 의한 해외여행

package tour : 교통, 숙박, 관광 등 모든 것이 포함된 전 일정 여행

chart tour : 전세여행

incentive tour : 포상여행

over land tour : 통과상륙여행(동일 국가 내에서 A 기항지에서 B 기항지로 이동하면서 통과상륙의 허가를 얻어 3~7일 정도의 여행을 하고 그 선박에 다시 승선하는 여행)

shore excursion : 기항지 상륙여행

Interest Tour : 고객의 관심을 유도하기 위하여 특별한 테마를 가지고 기획, 운영하는 여행

technical visit : 산업시찰여행

PVS(Passport, Visa, Shot-예방접종)

ASTA(American Society of Travel Agents) : 1931년 창립되어 미주에 본부를 두고 있는 미주 여행자 협회

WTO(World Tourism Organization) : 스페인 마드리드에 본부가 있고 정부 관광기관(우리나라는 문화관광부)이 회원으로 되어있는 세계 관광 기구

PATA(Pacific Asia Travel Association) : 1951년에 하와이 관광국의 제창으로 창립되고 샌프란시스코에 본부가 있는 태평양 아시아 관광 협회

2. Patterns Drill

① We'd like to see **Mt. Dobongsan.**
the interesting places
Mt. Maisan and Tapsa
the World Cup Stadium

② Do you know what that **building** is?
house
school
shop

③ Will you take my picture with a view of that **building**?
river
mountain
house
lake

④ I had a **wonderful** time there.
good
great
lot of fun

3. Dialogues

Dialogue 1

(G : Guide, T : Tourist)

G : Good morning. May I help you?
(안녕하세요, 무엇을 도와드릴까요?)

T : Yes, I'd like to see some city sights. Would you suggest some interesting places to visit in this city?

(예, 시내관광을 하고 싶습니다. 이 도시에서 구경할 만한 곳을 추천해 주시겠습니까?)

G : Sure. It's a sightseeing bus of the city. It'll take you by all the major points of interest.

(물론이지요. 시내 관광버스가 있습니다. 여행사에서 흥미 있는 모든 주요장소로 태워다 줄 것입니다.

T : How much is it?

(요금이 얼마입니까?)

G : It's $15 per person for a two- hour tour.

(2시간 관광에 1인당 15달러입니다.)

T : What time does the next tour leave?

(다음 관광은 몇 시에 출발합니까?)

G : It leaves at 10 a.m.

(10시에 출발합니다.)

Dialogue 2

G : What do you want to see here in Los Angeles.

(로스앤젤레스에서 무엇을 구경하고 싶으십니까?)

T : I don't know. I need some information. Would you please recommend some interesting places to visit?

(잘 모르겠습니다. 정보가 필요합니다. 구경할 만한 곳을 추천해 주시겠습니까?)

G : Sure. You should go to the Getty Museum. You can see a collection of LA's finest art.

(물론이지요. 게티 박물관을 가보십시오. LA의 가장 멋진 미술작품들을 구경할 수 있습니다.)

T : I can hardly wait to see it.

(그 곳을 빨리 구경하고 싶군요.)

G : The place is worth a visit.

(그 곳은 한번쯤 가볼만한 곳입니다.)

T : What time does the Getty Museum close?

(게티 박물관은 몇 시에 문을 닫습니까?)

G : I think it's closes at 6 o'clock.

(6시에 닫을 겁니다.)

T : Thank you for your information.

(정보를 주셔서 감사합니다.)

G : You're welcome.

(천만에요.)

4. Exercise

in bound

out bound

package tour

chart tour

incentive tour

over land tour

shore excursion

SIT(Special Interest Tour)

technical visit

PVS(Passport, Visa, Shot)

ASTA(American Society of Travel Agents)

WTO(World Tourism Organization)

PATA(Pacific Asia Travel Association)

07

3급 기관사 기관영어 문제

기관영어 문제

001 다음 일지 기재문의 밑줄 친 부분의 뜻은?

Found No.1 <u>plummer block</u> working hot.

가. intermediate shaft bearing
나. thrust shaft bearing
사. propeller shaft bearing
아. stern tube bearing

002 "공기 송출을 하기 전에 밸브를 잠그지 마시오."를 영문으로 표시할 때 (　　) 안에 알맞은 것은?

"Never (　　) your valve before air blowing"

가. lock
나. open
사. switch
아. restart

003 (　　) the joint surface of No.1 cylinder cover on M/E.

가. Faced up
나. Regulated
사. Screwed up
아. Tapped

해설 face up : 표면을 매끄럽게 하다.

004 S/B Eng. (　　) leaving port of Pusan.

가. to
나. for
사. of
아. on

005 Fuel oil tank vents are fitted with flame screens to prevent (　　).

가. escape of flammable vapors
나. flames entering the tank vent
사. corrosion in the tank vent
아. damage to the ball check

006 One of the important functions of lubricating oil circulation is ( ).

가. to cool the bearing caps by carrying away heat generated by friction.

나. to cool the bearing surfaces by carrying away heat generated by friction.

사. to increase the friction loss resulting from oil film.

아. to increase the revolution of shaft by lubrication.

해설 carry away : ~을 가져가버리다. 운반해가다. 빼앗다.

007 Most propellers on merchant vessels are made of ( ).

가. manganese bronze. 나. cast iron.

사. mild steel. 아. forged steel.

해설 manganese : 망간 / forged steel : 대장간 쇠, 제철소

008 Calipers are instruments that can be used to gauge the distance between two surfaces. 밑줄 친 부분과 뜻이 같은 것은?

가. create 나. enclose

사. fill 아. measure

009 Which of the following is a spark ignition engine?

가. gasoline engine 나. diesel engine

사. rocket engine 아. turbo-jet engine

010 Honing is ( ).

가. cutting process 나. grinding process

사. welding process 아. remetaling process

011 Renewed ( ) of clutch for No.2 F.O. purifier.

가. friction linings 나. spiral gear

사. gravity 아. separation disc

012 When the ship is at sea, the principal particulars to be entered in the log book are as follows : The (　　) of arrival and departure of the ship, the (　　) and details of the starting and stopping of the engines, and the (　　) and reason for reducing the speed.

가. time　　　　나. place
사. condition　　　　아. reason

013 When received fuel oil, which of the following specifications does not need to entry on engine log book?

가. price　　　　나. specific gravity
사. date, time　　　　아. quantity

해설 specifications : 명세, 내역, 상기

014 On the first page of the Log book, type of the (　　) and boiler other particulars are to be entered.

가. engine　　　　나. pump
사. steering gear　　　　아. capstan

015 "부산에 도착하였다."를 영문기관일지 기입법으로 맞게 표현한 것은 다음 중 어느 것인가?

가. Arrived Busan　　　　나. Arrived at Busan
사. Arrived for Busan　　　　아. Arrived of Busan

016 It is important that muster and drills should be carried out (　　), approaching as closely as possible to (　　) conditions.

가. fictionally, stationary　　　　나. realistically, emergency
사. fictionally, emergency　　　　아. realistically, stationary

해설 muster : 소집하다 검열 / drills : 훈련 / fictionally : 가상적으로 / realistically : 실제적으로

017 Before anyone is allowed to enter in the main engine crankcase, the (　　) should be engaged.

가. starting gear　　나. turning gear

사. cut-off servo　　아. governor

018 All the occurrences arising during the watch are to be entered engine log book by (　　).

가. chief engineer　　나. first engineer

사. the engineer on watch　　아. third engineer

해설 occurrence : 발생, 사건, 생긴 일

019 When the operating handle of a molded case circuit breaker is in mid-position, it indicates that the circuit breaker is (　　).

가. on　　나. off

사. tripped　　아. reset

020 To avoid excessive surge pressure when bunkering, you should (　　).

가. bunker at a nominal rate　　나. check for kinks in hose

사. close valves slowly　　아. shut valves before topping off

해설 kinks : 얽힘, 꼬임, 결함

021 See that (stand-by feed pumps) are ready for instant use. 괄호 부분의 뜻은?

가. 급수펌프　　나. 급수펌프에 의해 작동한다.

사. 주급수 펌프　　아. 예비급수펌프

022 (　　) means a person under training to become an engineer officer and designated as such by national law or regulations according to STCW convention.

가. Chief steward　　나. 2nd engineer

사. 3rd engineer

아. Assistant engineer

해설 designate : ~을 나타내다, 지시하다, 명시하다 / convention : 협정, 집회, 협의회 / assistant engineer : 실기사

023 All personnel on board should be fully ( ) of the precautions necessary for the ( ) of fire.

가. ignorant, prevention

나. aware, occurrence

사. ignorant, occurrence

아. aware, prevention

해설 ignorant : 무식한, 무지의 / prevention : 예방 / aware : 알고 있는, 감시한, 알아차린

024 Steering gear control must be provided both on the ( ) and in the steering gear room for the main steering gear.

가. bridge

나. engine room

사. master's room

아. chief engineer's room

025 ( ) 100 tons of F.O into No. 2 F.O.T.

가. Advanced

나. Renewed

사. Loaded

아. Worked

026 "출입금지"에 해당하는 용어는?

가. Way out

나. Not in use

사. Off limits

아. Entrance

해설 way out : 출구(exit)

027 When selecting the size of wire to be used in a circuit, the most important item to consider is the ( ).

가. size of the circuit

나. density of the circuit

사. amperage of the circuit

아. amount of wire to be used

028 If a boiler is allowed to run short of water, what action should be taken first?

가. close the uptake dampers

나. put out the fires

사. shut down the main stop valve

아. regulate the steam pressure by the hand lifting lever on the safety valve

029 The greatest care should be taken to guard against (   ) in steam piping by carefully draining all sections and other spaces where water might accumulate before (   ) is admitted.

가. steam leakage, drain

나. steam expansion, hot water

사. water hammer, steam

아. steam pressure, feed water

해설 guard : 보호하다 / water hammer : 수격작용

030 The solenoid valve can be typed as a (   ).

가. bellows valve

나. thermal valve

사. bimetallic valve

아. magnetic stop valve

031 The (   ) ships transport persons mainly.

가. merchant

나. passenger

사. cargo

아. time charter

해설 merchant : 상선

032 Which of the following is a compression ignition engine?

가. hot bulb engine

나. gasoline engine

사. diesel engine

아. steam turbine

033 When there is a Freon leak, the halide torch will burn (   ).

가. orange

나. blue

사. white　　　　아. green

해설 halide torch : 하알라이드 토치 / 녹색 : 소량누설 / 진한 청색 : 다량 누설

034 The purpose of (　　) rings are : to maintain a pressure tight seal between the piston and cylinder wall ; to aid in controlling oil ; to permit proper lubrication of the cylinders and to assist in cooling the piston.

가. oil　　　　나. lantern

사. mouth　　　　아. piston

035 S/B Eng. for arriving (　　) Busan.

가. on　　　　나. at

사. with　　　　아. to

036 다음 문장 중 ETD의 뜻은?

> How many days do you need for landing permit and what's your ETD?

가. estimated time of disembark

나. estimated time of departure

사. examine time of dock

아. examine time of document

해설 estimate : 추측의 예상의

037 Stopped cargo work (　　) to winch trouble.

가. in　　　　나. up

사. due　　　　아. back

해설 due to : ~ 때문에

038 The density of sea water depends on (　　).

가. temperature, pressure and salt content
나. temperature and surface tension
사. salt content, cohesion and viscosity
아. electrolytic content

039 If carbon dioxide has been discharged to extinguish or prevent a fire, the ( ) content will be reduced in the affected space.

가. hydrocarbon
나. hydrogen
사. carbon dioxide
아. oxygen

해설 carbon dioxide : 이산화탄소(CO_2) / affected space : 재난지역, 화재현장 / hydrocarbon : 탄화수소 / hydrogen : 수소 / oxygen : 산소

040 When you enter the pump room to rescue an unconscious person, which equipment must you have?

가. Protective clothing and explosion-proof flash light
나. Flame safety lamp and resuscitation equipment
사. Combustible gas indicator and cannister gas mask
아. Self-contained breathing apparatus and lifelines

해설 unconscious : 의식불명의 / resuscitation : 소생, 의식회복 / breathing apparatus : 호흡구

041 The ( ) is usually fitted on the suction side of a pump and must be cleaned regularly.

가. air chamber
나. strainer
사. coupling
아. gear

042 The engineer of the watch is never to absent himself from the ( ) during his watch on any account.

가. engine room
나. steering
사. bridge
아. control room

043 The combustible gas indicator(or explosimeter) detects the presence and proportion of (　　) vapour in air.

가. nitrogen　　나. oxygen
사. hydrocarbon　　아. hydrogen sulfide

해설 nitrogen : 질소 / hydrocarbon : 탄화수소 / hydrogen sulfide : 수소황화물

044 No person should enter boiler until it has (　　) sufficiently to make work safely in such place.

가. warmed up　　나. choked
사. illuminated　　아. cooled

해설 choke : 질식하다. 숨이 막히다. / illuminate : 비추다. 밝게 하다.

045 Each person's duties at fire drills and emergency drills are (　　).

가. in the engine room log book　　나. on the station bill
사. in the deck log book　　아. on the certificate of inspection

046 Before anyone is allowed to enter the main engine crankcase, the (　　) should be engaged.

가. starting gear　　나. turning gear
사. cut-off servo　　아. governor

047 "Overhauled and cleaned separation disc in L.O (　　)."

가. cooler　　나. purifier
사. tank　　아. heater

048 Engine horse power results from the engine's (　　).

가. speed and displacement
나. torque and speed

사. torque and compression ratio

아. compression ratio and displacement

049 The part of the crankshaft to which the connecting rod is attached is called the (　　).

가. journal　　나. throw

사. crank arm　　아. crank pin

050 (Pumped sufficient water) in No.4(P) tank to make the ship upright. What does (Pumped sufficient water) mean?

가. Filled up with water　　나. Ran down

사. Pumped out　　아. Loaded

051 In a refrigerating system, the purpose of the expansion valve by-pass is to (　　).

가. increase the efficiency of the plant

나. increase the capacity of the evaporator

사. control the refrigerant to the evaporator in case the automatic valves fail

아. by-pass the compressor

052 The sea water pump draws water from the (　　) through a strainer and delivers it through the engine fresh water cooler.

가. sea chest　　나. hull

사. filter　　아. ballast tank

053 The condensate water from coils in the fuel oil heaters goes to the (　　).

가. observation tank　　나. feed water heater

사. bilges　　아. auxiliary exhaust system

해설 condensate water : 응축수

054 The device used for the retention or quenching of sparks in exhaust pipes of internal combustion engines is called a (　　).

가. loop seal　　나. carburetor drip pan

사. spark arrester　　아. silencer

해설 retention : 포집, 보유, 보존, 억류 / quenching : 억누르다, 억제하다, (불을)끄다. / spark arrester : 그을음 제거기

055 The bulb for the thermal expansion valve in the refrigerating system is located (　　).

가. near the evaporator coil outlet

나. in the middle of evaporator coils

사. near the evaporator coil inlet

아. on the bottom row of evaporator coils

056 The tool used to cut threads in a hole is called a (　　).

가. top　　나. reamer

사. tap　　아. bit

057 Maintaining the atmosphere in any part of any cargo tank for tanker with an oxygen content not exceeding (　　) by volume.

가. 2[%]　　나. 5[%]

사. 8[%]　　아. 10[%]

058 Hydraulic test of piston to be (　　) after repair.

가. applied　　나. recoiled

사. started　　아. overhauled

059 (The engineer of the watch) is never to absent himself from the engine room. 에서

(　　) 부분에 속하지 않는 사람은?

가. 1st engineer　　나. 2nd engineer

사. 3rd engineer　　아. oiler

060 The Log Book is to be (　　) in a convenient place in the engine room.

가. equalized　　나. kept

사. healed　　아. penetrated

해설 equalize : ~을 똑같게 하다. 평등하게 하다. / penetrate : ~을 관통하다. 꿰뚫다. ~을 파악하다.

061 Some pistons are tapered with (　　) dia meter at the crown end.

가. the same　　나. the richer

사. the smaller　　아. the more

062 (　　) 안의 내용과 동일한 의미를 나타내는 것은?

> By removing one or more of these factors the fire will be (put out).

가. detected　　나. fired

사. extinguished　　아. postponed

해설 detect : 발견하다.

063 Compound gauge is necessary to read (　　).

가. both pressure and vacuum　　나. both pressure and temperature

사. only pressure　　아. only vacuum

064 All the occurrences arising during the watch are to be entered in the Log Book by (　　).

가. the engineer on watch　　나. third engineer

사. 1st engineer　　아. chief engineer

065 There are increased risks of electric shock to the welder if welding is done in hot or ( ) conditions.

가. dry
나. warm
사. cold
아. humid

해설 humid : 습기

066 The main causes of ( ) injury are infra-red rays, ultra-violet rays, exposure to chemicals, etc.

가. lung
나. kidney
사. eye
아. ear

해설 infra-red rays : 적외선 / ultra-violet rays : 자외선 / exposure : 노출 / lung : 폐, 허파 / kidney : 신장, 콩팥

067 Pointed out ( ) by PSC officer as follows;

> 1. Annual survey not conducted for safety construction cert.
> 2. The incinerator not working.

가. the efficiency
나. the flaw
사. the results
아. the deficiency

해설 point out : 지적하다. / PSC (port State Control) : 항만국통제 / Annual survey : 연차검사 / incinerator : 소각기(로) / efficiency : 효율, 능률 / flaw : 흠, 결점(도덕, 윤리상) / deficiency : 부족, 결핍, 결함

068 Fill blank(s) with suitable word(s).

> "( ) up B.W.T. with sea water."

가. Started
나. Filled
사. Full
아. Discharged

해설 fill up : ~을 채우다.

069 보조 보일러의 검사를 받기 위한 준비를 했다.

Made ( ) for ( ) of auxiliary boiler.

가. preparations, survey　　나. prepare, survey
사. preparations, dosing　　아. prepare, dosing

해설 preparation : 준비 / prepare : 준비하다 / survey : 검사 / dosing : 양을 복용하다

070 In a single stage impulse turbine the expansion of the steam takes place in ( ).

가. moving blade　　나. fixed blade
사. nozzle　　아. casing

071 The purpose of a ( ) is to keep the engine running at a desired speed regardless of the changes in the load carried by the engine.

가. governor　　나. overspeed trip
사. regulator　　아. flywheel

해설 regardless of : ~에도 불구하고, 개의치 않는

072 The mechanical efficiency of an engine is the ratio of the power actually delivered to the crankshaft called (①) to the power developed inside the cylinders called (②)

가. ① I.H.P, ② B.H.P.　　나. ① B.H.P, ② T.H.P.
사. ① B.H.P, ② S.H.P.　　아. ① B.H.P, ② I.H.P.

해설 mechanical efficiency : 기계효율

073 A synchroscope is used to ( ).

가. parallel AC generators
나. charge battery
사. synchronize propulsion motor and generator

아. connect breaker

해설 synchroscope : 동기검정기 / propulsion : 추진, 추진력

074 Nearly all fires can be easily controlled by one man with a portable ( ).

가. fire extinguisher
나. water tight door
사. look
아. fire detection

해설 nearly : 거의, 얼추

075 The pilot valve in a hydraulic governor ( ).

가. controls the oil pressure in the governor
나. controls the rate of injection of pilot oil where the governor is used on dual-fuel engines
사. controls the flow of oil both to and fro the power piston
아. controls only the flow of oil under pressure to the power piston

해설 to and fro : 앞뒤로 움직이는, 왕복의

076 When the ship is at sea, the principal particulars to be entered in the log book are as follows : The ( ) of water at the time of arrival and departure, weather, the direction and force of wind.

가. ton
나. quality
사. draft
아. ballast

해설 particular : 사항, 조목 / draft : 흘수

077 Took indicator cards for all cylinders of M/E to check ( ).

가. pressure gauge
나. bearing
사. combustion condition
아. crank arm deflection

078 Ventilation of battery room is required to ( ).

가. prevent accumulation of clean air

나. supply necessary inert gases for battery action

사. prevent accumulation of explosive gases

아. get moisture

079 The main purpose of the superheater vent valve on a boiler is to (　　).

가. insure a flow of steam through the superheaters when steam is being raised on the boiler

나. remove air from the superheaters

사. remove water from the superheaters

아. lower the steam pressure on a boiler that has just been secured

해설 insure : 안전하게하다. 보험에 들다. / superheater : 과열기

080 No engineer may (　　) either in a home or a foreign port without (　　) from the chief engineer.

가. go to ashore, permission　　나. go ashore, permission

사. go to ashore, prohibition　　아. go ashore, prohibition

해설 ashore : 물가로 / prohibition : 금지

081 A completed ship goes for (　　) before she is handed over to her new owners.

가. sea trials　　나. the maiden voyage

사. dry docking　　아. mooring

해설 sea trials : 시운전 / the maiden voyage : 처녀항해

082 Water in oil is easily separated when the oil has been (　　).

가. condensed　　나. freezed

사. cooled　　아. heated

해설 freeze-froze-frozen

083 Every lifebuoy shall be capable of supporting not less than 14.5[kgf] of iron in fresh water for a period of (   ) hours.

가. 12
나. 24
사. 48
아. 72

해설 not less than : 적어도

084 괄호의 단어와 같은 뜻의 단어는?

(Gauged) crank arm deflection of M/E.

가. Adjusted
나. Measured
사. Inspected
아. Charged

085 On a three-phase three-lamp ground detector, what would be the indication if one phase was grounded?

가. one lamp dark and the others bright.
나. one lamp bright and the others dark.
사. all the lamps would be dark.
아. all the lamps would be bright.

해설 ground detector : 접지검지기

086 "Measured air gap of induction motor for cooling pump."에서 밑줄 친 부분의 뜻은?

가. 공극
나. 공기벨트
사. 공기실
아. 진공실

해설 induction motor : 유도 모터

087 Discharged (   ) from main air tanks. Choose the most suitable one for the parenthesis.

가. air
나. oil

사. carbon residue　　아. drain

해설 residue : 나머지, 찌꺼기

088 The original (①) ratio of an engine is the volume of space when the piston is at BDC divided by the volume of combustion space when the piston is at TDC.

> Carbon deposits, which are incompressible solids, can accumulate in the combustion chamber and (②) the volume of combustion space, thereby increasing the ratio.

가. ① compression, ② reduce　　나. ① expansion, ② gain

사. ① explosion, ② increase　　아. ① specific heat, ② maintain

해설 thereby : 그것 때문에, 그것에 의하여

089 If two AC generators are to be operated in parallel, the load is distributed evenly by (　　).

가. a rheostat　　나. a balance coil

사. adjusting the governor settings　　아. changing excitation

해설 evenly : 균등히, 공평하게 / rheostat : 가감저항기 / excitation : (전기)여자

090 During engine design the firing order is determined by the (　　).

가. cylinder arrangement　　나. camshaft position

사. valve placement　　아. crankshaft design

해설 firing order : 폭발순서

091 The time which it takes to heat the fuel particles, turn into vapor, and start combustion is called ignition (　　).

가. advance　　나. time

사. lag　　아. period

해설 lag : 지연, 지체

092 The vertical distance that the piston travels between T.D.C. and B.D.C. is called the (   ).

가. bore
나. stroke
사. cylinder
아. cycle

093 (   ) is the play between the teeth of two gears which are engaged.

가. Gap
나. Back lash
사. Gear clearance
아. Chink

094 All life rafts to be opened and inspected, and replace the items whose term of validity are (   ).

가. begun
나. reserved
사. exploded
아. expired

095 (   ) is a cargo ship constructed or adapted for the carriage in bulk of liquid cargoes of an inflammable nature.

가. A bulk carrier
나. A tanker
사. A General cargo ship
아. A container

096 (   ) shall be completed on each occasion, on a tank-to-tank basis, whenever ballasting or cleaning of fuel oil tanks takes place in the ship.

가. Oil record book
나. Docking indent
사. Bell book
아. Abstract log

097 In accordance with the SOLAS 74/78 Convention as amended, the (   ) shall be capable of inerting empty cargo tanks by reducing the oxygen content of the atmosphere in each tank to a level at which combustion cannot be supported.

가. flame detector
나. oxygen analyzer

사. combustible gas detector
아. inert gas system

해설 amend : ~을 고치다. 변경하다.

098 All details of fuel bunkering, fuel oil tank cleaning or sludge disposal must be entered in the (   ).

가. rough log
나. engine room log
사. oil record book
아. abstract log

해설 disposal : 처리, 처분

099 We look forward (   ) hearing from you at your earliest convenience.

가. with
나. by
사. of
아. to

해설 at your earliest convenience : 형편대로, 되도록 일찍

100 The oil record book shall be preserved for a period of three years after the (   ) entry has been made.

가. first
나. last
사. beginning
아. start

101 A fire in its first few minutes can usually be readily (   ).

가. acceptable
나. distinguish
사. lighted up
아. extinguished

해설 readily : 즉시, 쉽사리

102 (   ) should enter a refrigerated compartment without first informing a responsible officer.

가. Every seaman
나. Anyone
사. No one
아. Chief steward

103 The watch from 0800 hours to noon is the (　　).

가. midnight watch　　나. morning watch

사. forenoon watch　　아. evening watch

해설 00~04 : midnight watch / 04~08 : morning watch / 08~12 : forenoon watch
12~16 : afternoon watch / 16~20 : evening watch / 20~24 : first night watch

104 Under the MARPOL 73 Convention as amended every (　　) of 150 gross tonnage and above shall be provided with an (　　).

가. passenger ship, oil record book

나. oil tanker, oil record book

사. passenger ship, engine log book

아. oil tanker, engine log book

해설 amended : 수정하다. 변경하다.

105 All life rafts to be opened and inspected, and replace the items whose term of validity are (　　).

가. begun　　나. reserved

사. exploded　　아. expired

해설 term of validity : 유효기간 / reserved : 예약하다. / exploded : 폭발하다. / expired : 만료되다. 끝나다.

106 Renewed 4 worn-out (　　) plates on condenser cover.

가. zinc　　나. stainless steel

사. asbestos　　아. canvas

해설 worn-out : 닳아 해진, 써서 낡은

107 "발전기 엔진이 300[rpm] 되도록 조속기 스프링을 조정했다."를 영문으로 표기할 때 (　　)에 알맞은 것은?

Adjusted (   ) spring of D/E to keep RPM 300.

가. fuel valve
나. governor
사. plunger
아. tappet

108 Changed Fuel oil for M/E (   ) MDO (   ) HFO.

가. to, with
나. from, to
사. from, by
아. to, of

109 If the discharge side of the thermal expansion valve is warmer than the inlet side, the (   ).

가. expansion valve is working normally
나. expansion valve is not working
사. solenoid valve is not working normally
아. scale trap is dirty

110 The sequence of an order in which power impulses occur in an engine is called the (   ).

가. expansion order
나. crank angle
사. ignition point
아. firing order

해설 firing order : 폭발순서

111 To ascertain the exact pressure of stroke of the piston is (   ).

가. spring
나. indicator
사. tachometer
아. manometer

해설 ascertain : ~을 확인하다. 규명하다.

112 If a boiler is found to run short of water, what action should be taken first?

가. close the uptake dampers.

나. put out the fires.

사. shut down the main stop valve.

아. regulate the steam pressure by the hand lifting lever on the safety valve.

해설 put out : 끄다.

113 In accordance with the SOLAS 74/78 Convention as amended, every steam boiler shall be provided with not less than two (　　) valves of adequate capacity.

가. safety　　나. stop

사. air vent　　아. check

해설 not less than : 적어도 / adequate : 충분한, 적절한

114 The disposal into the sea of food wastes is prohibited from the nearest land is less than (　　).

가. 3 nautical miles　　나. 12 nautical miles

사. 15 nautical miles　　아. 25 nautical miles

해설 nautical miles : 해리(1해리=1852[m])

115 Ships are less stable when water or fuel tanks are partially filled because of the (　　).

가. free-surface effect　　나. permeant sea water

사. increase in buoyancy　　아. leakage of liquid

해설 permeant : 침투하는, 스며드는

116 The difference between the pitch of the propeller and the actual advance of the ship in one revolution is called (　　). If the pitch is greater than the actual advance per revolution, this is positive ; if the pitch is less than the actual advance of the ship, this is said to be negative.

가. slip　　나. cavitation

사. gain　　아. speed

해설 gain : 이득

117 "기관사 호출장치는 적절하게 작동될 수 있도록 기관제어실 또는 기관 조종대 부근에 설치되어야 하며, 기관사 거주구역내에서 명확히 청취 가능해야 한다."를 영문으로 옮길 때 (　　) 속에 적합한 것은?

> An engineer's alarm shall be provided to (　　) from the engine control room or at the maneuvering platform as appropriate, and shall be clearly (　　) in the engineer's accommodation.

가. operate, audible　　나. operate, visible

사. be operated, audible　　아. be operated, visible

118 Overhaul lubricating oil pump for survey. 의 문장을 옳게 해석한 것은 어느 것인가?

가. 윤활유를 퍼내고 펌프를 검사하시오.

나. 펌프를 분해하고 윤활유를 검사하시오.

사. 수검을 위하여 윤활유 펌프를 개방 검사하시오.

아. 수검을 위하여 연료유 펌프를 개방 검사하시오.

119 No (　　) should be stored in the battery room.

가. hydrometer　　나. flammable material

사. hygrometer　　아. thermometer

해설 hydrometer : 비중계 / hygrometer : 습도계 / thermometer : 온도계

120 Which of the following is found on the surface of the plates when a lead-acid battery is discharged?

가. spongy lead　　나. lead sulphate

사. irons　　아. NaCl

해설 lead sulphate : 황산염

121 (　　) 20 tons of F.O from No. 2 tank to No. 3 tank.

가. Loaded　　　　나. Supplied

사. Shifted　　　　아. Received

122 Overhauled G.S. pump and renewed impeller shaft and (　　).

가. mouth ring　　　　나. oil ring

사. dam ring　　　　아. piston ring

123 Faced up (　　) valves of No.1 fuel pump on main engine.

가. suction and delivery　　　　나. main and auxiliary feed check

사. surface and bottom blow　　　　아. salinometer

해설 faced up : 연마하다.

124 Ventilation of the crankcase is provided in order to (　　).

가. aid in purifying the oil

나. maintain an uniform oil pressure

사. prevent the accumulation of combustible gases in the crank case

아. prevent the oil from becoming emulsified

해설 in order to : ~하기 위하여

125 In referring to threads, pitch is (　　).

가. the distance of the full length of the thread

나. the distance from a point on one thread to a corresponding point on the next thread measured parallel to the axis

사. the distance from the top of one thread to the bottom of the next thread

아. the distance from the bottom of a head on a bolt to the first thread

해설 refer : 언급하다. 알아보다. 참조하다.

126 The time between fuel injection and ignition in a diesel engine is known as (　　).

가. injection lag　　나. ignition lag

사. detonation lag　　아. precombustion lag

127 Which of the following is not ordinarily used for determining power factor?

가. voltmeter　　나. tachometer

사. ammeter　　아. wattmeter

128 Most coolers used on board ship transfer (　　) from a hot fluid to sea water.

가. pressure　　나. heat

사. weight　　아. load

129 A cofferdam is (　　).

가. an empty space between tank tops and bilges

나. a tank for storing chemicals

사. a cement dam in a fresh-water tank to prevent "sloshing"

아. an empty space separating compartments to prevent the contents of one compartment from entering another in case of leakage

해설 sloshing : ~(액체)가 출렁거리다.

130 After passing through the bearings and gears, the L.O. in a turbine installation goes to the (　　).

가. settling tank　　나. sump tank

사. daily tank　　아. storage tank

해설 installation : 설비, 장치

131 Clearance between (　　) and tail shaft to be measured.

가. trust bearing　　　　나. main bearing

사. stern tube　　　　아. connecting rod

해설 tail shaft : 선미축

132 Before reaching a home port should be ready a list of stores <u>on hand</u>. What is the meaning of underline?

가. remaining　　　　나. in hand

사. consumed　　　　아. within reach

133 While on the voyage, the engine must never be (　　) without timely warning to the captain or officer of watch, (　　) in cases of the most urgent necessity.

가. started, if　　　　나. started, unless

사. stopped, if　　　　아. stopped, unless

해설 timely : 때맞춘, 적시의 / in cases of : ~의 경우 / urgent : 긴박한, 급박한

134 All engineer officers and engine room ratings shall take (　　) to prevent such a pollution, particularly within framework of relevant international and port regulations.

가. appropriate precautions　　　　나. all possible precautions

사. cleaning　　　　아. appropriate effort

해설 rating : 부원 / framework : 구조물 / relevant : ~관련된

135 The basic requirement of deck machinery is to control (　　).

가. loads　　　　나. height

사. size　　　　아. width

136 The engine drives (　　) at the stern of a ship.

가. the rudder　　　　나. the propeller

사. the generator　　　　아. the motor

137 Petty officer in charge of the engine department is (　　).

가. No.1 oiler　　　　나. engineer officer

사. rating　　　　아. sailor

해설 Petty officer : 하급선원, (해군)하사관

138 Which of the following need not to be recorded in oil record book?

가. Date　　　　나. Port charge

사. Record of operations　　　　아. Signature of officer in charge

139 An "exciting current" is one that is required to (　　).

가. build up a dead circuit　　　　나. create a magnetic field

사. excite a induction motor　　　　아. build up the voltage in a battery

해설 exciting current : 여자전류 / magnetic field : 자장 / induction motor : 유도 모터

140 A steam pressure gauge is usually connected to a steam source by means of "a pig tail" or loop in order to (　　).

가. prevent steam from entering the gauge

나. prevent water from entering the gauge

사. provide for temperature compensation

아. provide for pipe expansion

해설 prevent : 예방하다. / provide : 제공하다. / compensation : 보상

141 The fuel ignites and burns during the first part of the working stroke, thus raising the pressure in the cylinder. During this stroke, called working or (①), the hot gases push the (②) downward.

가. ① explosion, ② cylinder　　　　나. ① impulse, ② crankshaft

사. ① power, ② piston　　　　아. ① expansion, ② crankshaft

142 An insulator is a ( ).

가. substance that offers a high resistance to current flow

나. substance that offers a low resistance to current flow

사. substance that absorbs electricity

아. form of a condenser

해설 insulator : 절연체, 절연물 / absorb : 흡수하다.

143 A heat engine is a machine for the conversion of heat energy into ( ), and its function must be to make the conversion as efficiently as possible.

가. reciprocating motion　　　　나. centrifugal force

사. mechanical energy　　　　아. resistance

144 The function of fuel oil ( ) is to prevent foreign matter from passing with oil to the burners.

가. stopper　　　　나. cock

사. valve　　　　아. strainer

해설 foreign matter : 이물질

145 Before synchronizing alternators, they must have ( ).

가. the load equalized　　　　나. the same voltage

사. the same horse power　　　　아. the same ampere

해설 synchronizing alternator : 동기발전기

146 An instrument used to measure high temperature is ( ).

가. manometer　　　　나. pyrometer

사. hydrometer　　　　아. thermometer

해설 manometer : 마노미터(유체압력계) / pyrometer : 고온계 / hydrometer : 비중계 / thermometer : 온도계

147 "한국선급의 검사원 J씨의 입회하에서 안전밸브의 자물쇠를 채웠다."를 영문으로 옮길 때 (　　)에 알맞은 것은?

(　　) the safety valve (　　) Mr.J, K.R. surveyor.

가. Closed up, at the presence of　　나. Locked up, in the presence of
사. Closed up, with the presence of　　아. Locked up, with the presence of

해설 in the presence of : ~의 면전에서

148 "The engineer is never to give over the charge of engine to any oiler."의 영문이 뜻하는 것은?

가. 당직기관사는 자유롭게 기관실을 비울 수 있다.
나. 당직기관사의 책임을 부원에게 맡겨서는 안 된다.
사. 기관장은 당직을 당직부원에게 맡길 수 없다.
아. 당직기관사의 책임을 기관장에게 위임할 수 있다.

149 An electric tachometer receives the engine speed signal from a/an (　　).

가. small generator mounted on the engine
나. bimetalic sensing device
사. stroboscopic sensing device
아. vibration sensing device

150 Which of the following instruments is not needed in the operation of 2 DC generators operating in parallel?

가. voltmeter　　나. ammeter
사. synchroscope　　아. over current relay

해설 synchroscope : 동기검정기

151 Steering gear control must be provided both on (　　) and in the steering gear room for the main steering gear.

가. the engine room　　나. the bridge

사. the deck　　아. the stern of ship

152 The oil record book shall be kept in such a place as to be readily available for (　　).

가. modification　　나. copying

사. inspection　　아. detection

해설 modification : 수정 / inspection : 검사 / detection : 발견

153 "(Took out) No.2 cylinder liner and repaired crack parts." What does (took out) mean?

가. Checked out　　나. Draw out

사. Tightened up　　아. Skimmed up

해설 draw out : 뽑아내다, 제거하다 / skimmed : ~을 걷어내다, 떠내다

154 (　　) F.O from port tank to starboard tank.

가. Shifted　　나. Replaced

사. Filled　　아. Drained

155 The (　　) point of an oil is the temperature at which inflammable vapours are given off in sufficient quantity to flash when brought into contact with a flame.

가. fire　　나. ignition

사. flash　　아. pour

156 Inspected all working parts in (　　) of M/E, found 1 crank pin bolt loosened, and retightened it up.

가. cylinder liner
나. crankcase
사. cylinder head
아. camshaft

157 Engine displacement can not be found unless the ( ) is known.

가. power ratio
나. compression ratio
사. bore and stroke
아. compression space

해설 engine displacement : 배제용적

158 The purpose of the flywheel is to ( ).

가. prevent over speeding of the engine
나. drive accessories
사. reduce the variations in crankshaft speed between power strokes
아. counter balance the crank throws on the crankshaft

해설 throws : ~을 던지다. ~을 움직이다.

159 The cetane number of a fuel sample is determined by testing it in a single cylinder test engine with ( ).

가. compression ignition
나. spark ignition
사. variable compression ratio
아. fixed compression ratio

160 The turning effort produced by the engine is called a ( ).

가. torque
나. power ratio
사. twist
아. axial force

161 ( ) is a property of fluids and is a measure of resistance to flow. It is determined in a practical sense by measuring the time it takes a given quantity acted on by gravity to pass through an orifice.

가. Specific gravity
나. Pour point

사. Viscosity 아. Heating value

해설 gravity : 중력 / specific gravity : 비중 / pour point : 유동점 / viscosity : 점도

162 Static electricity is most often produced by (   ).

가. pressure 나. friction

사. magnetism 아. heat

해설 static electricity : 정전기 / magnetism : 자기, 자력

163 A ring which keeps a person afloat in the water is (   ).

가. lifebuoy 나. lifeboat

사. life jacket 아. liferaft

164 (   ) Osaka for Busan.

가. Came 나. Left

사. Went 아. Arrived

165 The engineer of the (   ) is never to absent himself from the engine room.

가. oiler 나. first

사. watch 아. electrician

166 If you notice oil in the vicinity of the vessel while bunkering, you should (   ).

가. notify the master

나. stop bunkering

사. ascertain if oil is coming from your vessel

아. notify terminal superintendent

해설 vicinity : 근처 / ascertain : ~을 확인하다. 규정하다. / superintendent : 감독(자), 관리자

167 Before repair work is started on a dynamo engine, it should be (   ) and the starting air valve should be secured so that it cannot be operated.

가. slow-down　　나. run idle

사. stopped　　아. turned

해설 secured : ~을 안전하게 하다.

168 A person who is in command and charge of a whole ship is ( ).

가. the master　　나. the chief officer

사. the oiler　　아. the chief engineer

169 Air starting systems for two-stroke cycle engine require a minimum of ( ) cylinders in the engine. This is because the cycle is 360° rather than the 720° of the four-stroke cycle.

가. two　　나. three

사. four　　아. six

해설 rather than : ~보다 오히려

170 "다시 죄도록 하시오."를 영문으로 옮길 때 ( )속에 적합한 용어는?

( ) it again.

가. Fasten　　나. Unite

사. Feed　　아. Unbind

해설 fasten : ~을 단단히 고정시키다. 잠그다. 죄다. / unite : ~을 하나로 하다. 결합하다.

171 Choose the most suitable one for the blanks.

Measured top clearance of all pistons with ( ).

가. steel　　나. copper

사. lead　　아. iron

172 Instantaneous rate of discharge of oil content means the rate of discharge of oil in ( ) per ( ) at any instant divided by the speed of the ship in ( ) at the same instant.

가. liters, hour, knots
나. liters, minute, knots
사. cubic meters, hour, knots
아. cubic meters, minute, knots

해설 Instantaneous : 즉시의, 순간의 / Instantaneous rate of discharge : 순간배출율 / instant : 잠깐, 즉시, 순간

173 Choose the most suitable one for the blanks.

S/B Eng. ( ) leaving port of Busan.

가. to
나. for
사. of
아. on

174 Choose the most suitable one for the blanks.

The total resistance of a parallel circuit is always ( ).

가. larger than that of the branch with greatest resistance
나. smaller than that of the branch with the lowest resistance
사. equal to the sum of the individual branch resistance
아. equal to the reciprocal of the sum of the individual branch resistance

175 Choose the most suitable one for the blanks.

In six cylinder and four stroke cycle engine, there are six power strokes for every ( ) revolution of the crank shaft.

가. three
나. two
사. six
아. four

176 Choose the most suitable one for the blanks.

> The entrainment of a considerable portion of moisture by the steam leaving the boiler is called (   ).

가. steam bubble
나. priming
사. foaming
아. saturated steam

해설 entrainment : 함유되는, 포함되는 / moisture : 수분

177 Shims are used to (   ).

가. replace fuses
나. separate fresh and sea water lines
사. adjust cylinder liners
아. adjust bearing clearances

178 In two stroke diesel engine most efficient cylinder lubrication is obtained by injecting the oil into the piston, at the beginning of the (   ).

가. compression stroke
나. suction stroke
사. working stroke
아. exhaust stroke

179 Measured clearance between stern tube bearing and (   ).

가. propeller shaft
나. crank shaft
사. intermediate shaft
아. impeller shaft

180 "Winch to be opened up for survey." 에서 opened up의 뜻은?

가. 개방하다.
나. 폐쇄하다.
사. 수리하다.
아. 연기하다

181 The engine log book is to be kept in a convenient place in (   ).

가. engine room
나. galley
사. bridge
아. cargo control room

해설 galley : 요리실, 취사장

182 Officers in charge of an engineering watch shall keep the main propulsion plant and auxiliary system under constant supervision until properly (　　), and shall periodically (　　) the machinery in their charge.

가. common, survey　　나. relevant, inspect
사. relieved, inspect　　아. watch, survey

해설 supervision : 감독, 관리 / periodically : 주기적으로 / charge : 임무

183 This is rule for dealing with and writing up the engine room log book, that is, when a (　　) has been made in the Log Book, two black lines shall be drawn through that part.

가. particulars　　나. watch
사. misentry　　아. Chief Engineer

184 The (　　) shall be preserved for a period of three years after the last entry has been made.

가. engine maintenance log　　나. oil record book
사. bell book　　아. abstract log

해설 preserved : ~을 보존하다, 보관하다

185 Officers in charge of an engineering watch, as defined in the STCW Code and under the direction of the chief engineer officer, shall be immediately available and on call to (　　) the machinery spaces.

가. attend　　나. absent
사. stand　　아. keep

해설 in charge of : ~을 맡아서, 담당해서

186 Which of the following would not cause a centrifugal type pump to fail to discharge liquid?

가. pump not primed
나. insufficient speed
사. impeller plugged or broken
아. suction lift too low

해설 suction lift : 흡입양정

187 "알파라발 유청정기에 관한 최신 정보가 게재된 회보를 보내 주시기 바랍니다."를 영문으로 옮길 때 ( )에 알맞은 것은?

Please send ( ) containing ( ) information concerning the Alfa Laval Oil Purifier.

가. bulletins, the latest
나. bulletins, the last
사. labels, the latest
아. labels, the last

해설 bulletins : 편람

188 Fill in the blank with suitable words or phrases.

Filled up ( ) with sea water.

가. settling tank
나. air tank
사. ballast water tank
아. sump tank

189 A relief valve can be loaded with a spring so that it may open automatically when ( ) is too high.

가. gas
나. temperature
사. pressure
아. electricity

190 The intake and exhaust valves are closed and the air in the cylinder is compressed in the ( ) resulting in high temperature.

가. suction stroke
나. expansion stroke

사. compression stroke　　　　아. exhaust stroke

191 If an engine could convert all the heat developed by burning of the fuel into useful work, its (　　) would be 100%.

가. heat ratio　　　　나. thermal efficiency

사. mechanical efficiency　　　　아. thermal ratio

192 Took (　　) all accessories of No.4 cylinder cover for M/E.

가. in　　　　나. out

사. of　　　　아. up

해설 take in : ~을 받아들이다. 마시다. / take out : ~을 꺼내다. / take up : ~을 들어 올리다.

193 If the safety valve starts to whistle on an oxygen-breathing apparatus(OBA), the person wearing the OBA should (　　).

가. go out into the fresh air

나. open the by-pass valve wide

사. close the by-pass valve until the whistling stops

아. reset the timer for an additional 10 minutes

해설 oxygen-breathing apparatus : 산소구명장치

194 The watch engineer should not leave the watch (　　) until properly relieved.

가. time　　　　나. position

사. during　　　　아. engine

195 All air receivers and other pressure vessels for essential services together with their mountings and safety devices are to be opened up and examined internally and externally. If internal examination of them is not practicable, they are to be tested hydraulically to (　　) the working pressure in accordance with the KR class rules.

가. 1.1 times　　나. 1.2 times
사. 1.25 times　　아. 1.5 times

196 Batteries should always be transported in the (　　) position to avoid spillage of electrolyte.

가. reverse　　나. upside down
사. upright　　아. declined

해설 declined : ~을 거절하다, 사절하다.

197 "어느 구역의 침수율이라 함은 그 구역에 있어서 물이 차지할 수 있는 용적과 그 구역의 전용적과의 비율을 말한다."를 영문으로 옮길 때 (　　)에 알맞은 것은?

(　　) of a space means the ratio of the volume within that space which assumed to (　　) by water to the total volume of that space.

가. Broken space, be occupied　　나. Broken space, occupy
사. Permeability, be occupied　　아. Permeability, occupy

해설 permeability : 침투성, 투과성

198 (　　) is a machine for the conversion of heat energy into mechanical energy, and its functions must be to make the conversion as efficiently as possible.

가. An electric motor　　나. A ref. machine
사. A pump　　아. A heat engine

해설 heat engine : 열기관

199 A "Hydrometer" is used to measure (　　).

가. pressure　　나. pH
사. specific gravity　　아. temperature

해설 hydrometer : 비중계 / specific gravity : 비중

200 A period of duty of a ship's crew is (　　).

가. watch　　나. master

사. sailor　　아. first officer

201 The (　　) is responsible for the engine department.

가. master　　나. chief engineer

사. boatswain　　아. watch officer

202 Adjusted (　　) of No.2 crank pin bearing by putting in 2 sheets of 1 mm copper liner and by taking out 1 sheet of 2 mm brass liner.

가. oil temperature　　나. oil pressure

사. oil clearance　　아. oil leakage

해설 putting in : 끼워 넓 / taking out : 빼다.

203 "Opened plugs and (blew down) all water of B.W.T."에서 괄호 부분과 같은 뜻은?

가. charged　　나. discharged

사. loaded　　아. stopped

해설 B.W.T. ballast water tank

204 Opened test cocks of all cylinders for M/E and carried (　　) air running.

가. to　　나. with

사. out　　아. on

해설 carry with : ~을 휴대하다. / carry out : ~을 수행하다. / carry on : 경영하다. 관리하다.

205 Which is an incorrect simpler form?

가. Slow astern : S/A　　나. Motor Ship : M/S

사. Fuel Oil Tank : F.O.T　　아. Finished with engine : F.W.E

해설 simpler form : 약어

206 The crankshaft is of alloyed steel forging. To give better dynamic balancing, each ( ) is fitted with a counter-weight.

가. lever
나. web
사. journal
아. crank pin

해설 alloyed steel : 합금강 / forging : 쇠를 벼려서 만들다. / counter-weight : 평형추 / crank arm= crank web

207 The thermal expansion valve of refrigerator responds to the ( ).

가. amount of superheat in the vapor leaving the coil
나. amount of superheat in the liquid
사. temperature in the evaporator coils
아. pressure in the evaporator coils

해설 responds : 반응하다

208 Engaged turning gear and turned main engine for 20 minutes after F.W.E.

What does the underlined F.W.E. stand for?

가. Finish Within Engine
나. Finish With Engineer
사. Finished With Engineering
아. Finished With Engine

해설 engage : 작동시키다.

209 Choose the most suitable one for the blank.

All crew should be careful to ( ) cigarettes completely in their cabin.

가. protect
나. extinguish
사. build up
아. load

210 When received fuel oil, which of the following specifications does not need to entry on engine log book?

가. price of fuel oil
나. specific gravity
사. viscosity
아. flash point

해설 specific gravity : 비중 / viscosity : 점도

211 "Opened both end covers of L.O cooler for M/E and cleaned sea water side. Plugged 2 pieces of defective tube of jacket F.W cooler" What is explained to inspection of the auxiliary machine?

가. air conditioner
나. oily water separator
사. fresh water generator
아. heat exchanger

해설 defective : 결함이 있는

212 '조기수 김씨가 오늘 승선하였다'를 영문기관일지에 바르게 기입한 것은?

가. Oiler, Kim joined her today.
나. Wiper, Kim joined her today.
사. Oiler, Kim landed her today.
아. Wiper, Kim onboard her today.

213 Emergency generators must be located (　　).

가. in any dry space
나. above bulkhead or freeboard deck
사. close to the engine room
아. near the bridge

해설 bulkhead : 격벽 / freeboard 건현

214 When the frequency of the power stroke impulses is the same as the natural rate of torsional vibrations of the crankshaft, the shaft is then said to be rotating at a (　　).

가. rich over
나. sea speed
사. critical speed
아. normal speed

해설 natural rate of torsional vibrations : 고유진동수 / critical speed : 위험회전수

215 The engineer officer on watch shall notify (　　) immediately in the event of fire.

가. the fire station
나. administration

사. the bridge　　아. chief engineer

해설 immediately : 즉시 / administration : 관리, 지배

216 A "salinometer" indicates (　　).

가. vacuum　　나. pH

사. density　　아. the proportion of salt

해설 salinometer : 검염계

217 The space between the piston at TDC and the cylinder head is the (　　).

가. clearance volume　　나. combustion volume

사. stroke volume　　아. total volume

218 The purpose of (　　) rings are : to maintain a pressure tight seal between the piston and cylinder wall ; to aid in controlling oil ; to permit proper lubrication of the cylinders and to assist in cooling the piston.

가. oil　　나. lantern

사. mouth　　아. piston

해설 permit : 허락하다.

219 The oil record book shall be preserved for a period of (　　) years after the last entry has been made.

가. one　　나. two

사. three　　아. five

220 "부산에 입항하기 위해 기관을 스탠바이 하였다"를 영문 기관일지 기입법으로 맞게 표현한 것은?

가. S/B Eng. to arriving Busan.　　나. S/B Eng. to arriving at Busan.

사. S/B Eng. for arriving Busan.　　아. S/B Eng. for arriving at Busan.

221 The viscosity of diesel fuel (　　).

가. is a measure of its resistance to flow

나. has no affect on the nozzle spray pattern

사. is measured with a hydrometer

아. determines the heating valve of the fuel

해설 hydrometer : 비중계

222 Before drilling a hole in a piece of metal, it should be (　　).

가. center-punched　　나. marked with chalk

사. scribed　　아. protracted

해설 chalk : 분필 / protracted : ~을 연장하다.

223 When something goes wrong in an automatic plant.

가. an alarm flash

나. a light will sounds

사. the fires will go out

아. an alarm sounds and an indicating light will flash

224 Decreased F.H.N. from 9 to 6 slowly for stormy weather.

What does the underlined F.H.N. stand for?

가. Fuel Hand Number　　나. Fuel Handle Notch

사. Fuel Have Number　　아. Fuel Have Notch

해설 stormy : 폭풍우의

225 On receiving the order "Finished With Engine" the main air stop valve should be (　　) ; all the air drains should be opened ; the auxiliary scavenge air blower should be (　　).

가. opened, run　　나. opened, stopped
사. closed, run　　아. closed, stopped

226 Which one of the following is not one of six basic parts of the business letters?

가. letter head　　나. signature
사. salutation　　아. remarks

해설 letter head : 머리말 / salutation : 인사(말)

227 The engineer of the watch is (　　) to absent himself from the engine room, during his watch, on any account.

가. usually　　나. sometimes
사. used　　아. never

228 (　　) is the vertical distance from the moulded base line amidships to the subdivision load line in question in accordance with the provision of the SOLAS Convention.

가. Freeboard　　나. Breadth
사. Draft　　아. Length

해설 amidships : 배의 중앙에서, 배 중앙부에 있는 / draft : 흘수

229 Leakage of the Freon gas may be detected by the use of (　　).

가. halide torch　　나. litmus paper
사. filter　　아. alkali

230 Fuel oil tanks to be (　　) and cleaned out for internal inspection by surveyor.

가. faced up　　나. freed from gas
사. polished　　아. ground off

해설 cleaned out : 청소하다, 정돈하다 / polished : ~을 닦다, 윤이 나다

231 A life jacket shall have buoyancy which is not reduced by more than 5% after ( ) hours submersion in fresh water.

가. 12　　나. 24

사. 48　　아. 72

해설 buoyancy : 부력, 부양 / submersion : 침수, 잠수

232 According to the SOLAS Convention as amended, immersion suit is a protective suit which reduces the ( ) of a person wearing it in cold water.

가. body humidity　　나. mental consciousness

사. brilliance of skin　　아. body heat-loss

해설 immersion suit : 잠수복

233 "검사에 대비하여 선미축을 빼어 낼 것."을 영문으로 옮길 때 ( )에 알맞은 것은?

Tail shaft ( ) for survey.

가. to be drew out　　나. to draw out

사. to be drawn out　　아. to drew out

해설 draw out : 뽑아내다, 제거하다.

234 Cut off No.6 fuel ( ) on main engine.

가. injection pump　　나. feed pump

사. booster pump　　아. transfer pump

해설 cut off : ~을 잘라내다, 차단하다

235 Generator, electric motors, lighting circuits, etc. to be ( ).

가. megger tested　　나. pressure tested

사. running tested　　아. on line tested

236 A distiller condenses (　　) produced by the evaporator.

가. condensate　　나. water

사. vapor　　아. drain

해설 distiller : 증류수(기) / condense : 응축하다.

237 A supercharger (　　).

가. is a type of air pump which takes air out of the atmosphere, compresses it, and feeds it into the engine.

나. provides a greater volume of air at atmospheric pressure

사. is only of a positive displacement type

아. is always driven by belt or chain

238 Which of the following is not the internal combustion engine?

가. diesel engine　　나. gas turbine engine

사. steam turbine engine　　아. gasoline engine

239 The tool used to check external pipe diameter is called a (　　).

가. plug gauge　　나. pitch gauge

사. thread gauge　　아. ring gauge

해설 ring gauge : 플러그게이지와 한조로 되어있는 고리모양의 게이지

240 All details of fuel bunkering, ballasting or cleaning of fuel oil tanks, collection and disposal of oil residues must be entered in the (　　).

가. oil record book　　나. ship's official log book

사. noon report　　아. docking indent

해설 residue : 나머지, 찌꺼기 / indent : 주문

241 Which of the followings is not used for complimentary close of letter?

가. Yours truly
나. Respectfully yours
사. My dear Mr. Smith
아. Sincerely yours

해설 complimentary : 서한의 말미에 사용하는 경어 / Respectfully : 정중하게 / Sincerely : 진심으로

242 According to SOLAS convention, a passenger ship is a ship which carries more than ( ) passengers and cargo ship is any ship which is not a passenger ship.

가. twelve
나. thirteen
사. fifteen
아. twenty

243 "부산을 출항하기 위해 기관을 스탠바이 하였다"를 영문으로 기관일지에 기입할 때 맞게 표현한 것은?

가. S/B Eng' in leaving Busan.
나. S/B Eng' in leaving from Busan.
사. S/B Eng' for leaving Busan.
아. S/B Eng' for leaving from Busan.

244 Inspected and cleaned ( ) of No.1 air cooler for M/E.

가. water and oil side
나. air and oil side
사. oil and chemical side
아. water and air side

245 The packing in the gland must be kept soft and in good condition to prevent the ( ) of air and fresh water.

가. erosion
나. settlement
사. keep
아. escape

해설 erosion : 부식 / settlement : 정착

246 In motor ships there are two circulating water systems, one sea water, the other ( ) water.

가. salt
나. hot
사. fresh
아. ice

247 The lime in cylinder jackets, if deposited as scale, acts as (　　) of heat.

가. transfer　　나. conductor

사. transmission　　아. non-conductor

해설 lime : 석회

248 The auxiliary exhaust steam is used to (　　).

가. heat the cargo oil　　나. heat the feed water

사. drive the feed water pump　　아. drive the turbo generator

249 In a diesel engine, blow-by (　　).

가. increases exhaust back pressure　　나. increases liner and piston ring wear

사. decreases fuel consumption　　아. increases engine output

250 When a/an (　　) has been made in the log book, a black line shall be drawn through that part.

가. entry　　나. misentry

사. record　　아. report

251 (　　) means the senior engineer officer responsible for the mechanical propulsion and the operation and maintenance of the mechanical and electrical installations of the ship.

가. Chief engineer　　나. 1st engineer

사. 2nd engineer　　아. 3rd engineer

252 다음은 무엇을 설명한 것인가?

> A strong boat for saving persons after an accident at sea.

가. life jacket　　나. life boat

사. lift ling　　　　아. soap

253 (　　) of the log book, the type of the engines and boilers and other particulars are to be entered.

가. On the last page　　　　나. On the second page

사. On the cover　　　　아. On the first page

254 K사로부터 1,500 톤의 중유를 수급했다.

(　　) from K oil Co., LTD.

가. Received light fuel oil of 1,500 M/T

나. Received 1,500 M/T of heavy fuel oil

사. Receipt light fuel oil of 1,500 M/T

아. Receipt 1,500 M/T of heavy fuel oil

255 If the high-pressure switch on the compressor of a refrigerator opens frequently, a possible cause could be (　　).

가. a leak in the evaporator coils

나. not enough cooling water going through the condenser

사. too much cooling water going through the condenser

아. not enough refrigerant in the system

256 Heat added to a substance without causing a change of state will cause a change of temperature. This heat is called a (　　) because its addition is perceptible on the thermometer.

가. latent heat　　　　나. radiant heat

사. specific heat　　　　아. sensible heat

해설 change of state : 상변화 / perceptible : 지각(감지)할 수 있는 / latent heat : 잠열 / radiant heat : 복사열 / specific heat : 비열 / sensible heat : 현열

257 The volume swept by the piston from bottom dead center to top dead center is called ( ).

가. compression ratio　　나. power ratio

사. allowable compression ratio　　아. piston displacement

258 ( ) the camshaft speed is equal to half that of the main shaft.

가. In 4 stroke engine　　나. In the modern rotary engine

사. In 2 stroke engine　　아. In the steam reciprocating engine

259 Lignumvitae is a hard, dark-colour ( ) that does not swell, and is a water-lubricated bearing surface because it is always under water.

가. wood　　나. plastic

사. rubber　　아. synthetic

해설 swell : 부풀다. 팽창하다. / synthetic : 인조

260 The ( ) point is the temperature at which the oil will continue to burn when ignited.

가. dew　　나. explosion

사. fire　　아. ignition

해설 dew point : 이슬점

261 Stopped cargo work ( ) to winch trouble.

가. in　　나. up

사. due　　아. back

262 What is the temperature at which volatiles will ignite momentarily at the surface on the passing of a small flame.

가. freezing point　　나. boiling point

사. fire point

아. flash point

해설 volatile : 휘발성의, 증발하기 쉬운

263 What does STCW stand for?

가. Suffering of Training, Certification and Working

나. Standards of Training, Certification and Watch keeping

사. Standing of Training, Certification Watch keeping

아. System of Training, Certification and Working

264 "주기관의 회전수를 서서히 증가시켜 10시에 핸들 위치를 80에 설정했다."를 영문으로 옮길 때 (　　)에 알맞은 것은?

> Increased revolution of main engine (　　) and (　　) handle position to 80 at 10 : 00.

가. step in step, set

나. step by step, set

사. step by step, setted

아. step in step, setted

해설 step by step : 한 걸음 한 걸음, 점차로, 착실히

265 Refilled No.3 F.W.T. (　　) fresh water.

가. to

나. for

사. with

아. by

266 In its flow path through a diesel engine, which of the following internal parts does the lubricating oil system oil usually reach first?

가. piston pin bearing

나. cam shaft bearing

사. main bearing

아. cylinder walls

267 The purpose of the evaporator in a refrigerating system is to (　　).

가. absorb latent heat of vaporization　　나. absorb latent heat of fusion
사. transfer latent heat of vaporization　　아. transfer latent heat of fusion

268 The space between the piston at TDC and the cylinder head is the ( ).

가. clearance volume　　나. combustion volume
사. stroke volume　　아. total volume

269 "육안으로 볼 수 있는 범위 내에서는 추진기의 끝 부분이 경미한 손상을 입었음을 알았습니다."를 영문으로 옮기면 ( )에 알맞은 것은?

I found the propeller tip was damaged slightly ( ) my ( ) could reach.

가. so far as, eye sight　　나. as good as, eye sight
사. so far as, sightseeing　　아. so good as, sightseeing

해설 so far as : ~하는 한, 거기까지 / as good as : ~과 같은, ~에 충실한 / eye sight : 시력 / sightseeing : 관광구경, 유람

270 Before going off duty, the officer in charge of the engineering watch shall ensure that all events related to the main and auxiliary machinery which have occurred during the engineering watch are suitably ( ).

가. surveyed　　나. recorded
사. relieved　　아. seen

271 "Any engine order given should be <u>repeated</u> by the person operating the bridge telegraph"에서 밑줄 친 부분의 뜻은?

가. 확인해야 한다.　　나. 복창해야 한다.
사. 점검해야 한다.　　아. 금지해야 한다.

272 According to the ( ) convention, every oil tanker of 150 tons gross tonnage and above shall be provided with an oil record book.

가. MARPOL　　　나. IMO

사. STCW　　　아. SOLAS

273 Adjusted (　　) of No.2 crank pin bearing by putting in 2 sheets of 1[mm] copper liner and by taking out 1 sheet of 2[mm] brass liner.

가. oil temperature　　　나. oil pressure

사. oil clearance　　　아. oil leakage

274 "천천히 최대로 올리시오."를 영문으로 옮길 때 (　　) 속에 적합한 것은?

"Turn it up to (　　) slowly."

가. full　　　나. blank

사. low　　　아. stop

275 A generator has been exposed to water. Which of the following instruments would be used to determine if it is safe to operate?

가. a megger　　　나. a voltmeter

사. a hydrometer　　　아. a wattmeter

276 Seafarers required to work in machinery spaces which have high noise levels should (　　) suitable hearing protection.

가. discard　　　나. take off

사. wear　　　아. remove

해설 seafarers : 선원 / discard : ~을 버리다. 포기하다. / take off : ~을 제거하다. 없애다.
wear : 착용하다.

277 The term "engineering watch" means a period of (　　) for an officer during which the physical presence in machinery spaces of that officer may or may not be required.

가. watch　　　　나. operation

사. responsibility　　　　아. voyage

해설 responsibility : 책임, 의무

278 Before going off duty, the officer in charge of the engineering watch shall ensure that all events related to the main and auxiliary machinery which have occurred during the engineering watch are suitably (　　).

가. surveyed　　　　나. recorded

사. relieved　　　　아. seen

해설 in charge of : ~을 맡은 / ensure : ~을 책임지다, 보증하다.

279 Who is responsible for the efficient operation of the main engine and all machinery on board?

가. master　　　　나. chief engineer

사. chief officer　　　　아. first engineer

280 The watch engineer is never to absent himself from the engine room (　　) his watch.

가. in　　　　나. on

사. during　　　　아. at

281 (　　) system is used to keep the ship trimmed.

가. The ballast　　　　나. The heating

사. The mooring　　　　아. The sewage

282 Which of the following need not to be recorded in oil record book?

가. Date　　　　나. Sea chest in use

사. Record of operations　　　　아. Signature of officer in charge

283 A (　　) is a speed sensitive device used to control or limit the engine speed.

가. speed log　　　　나. tachometer

사. load limit　　　　아. governor

284 The charging of a storage battery at too high rate causes an excessive formation of gas bubbles which tends to wash away the active material from the (　　) plate.

가. negative　　　　나. positive

사. emergency　　　　아. top

해설 bubbles : 거품 / wash away : 씻어 내리다.

285 The lime, if deposited as scale, acts as nonconductor of heat and (　　) considerably the transfer of heat through the walls of the cylinder liner.

가. reduces　　　　나. occurs

사. increases　　　　아. allows

286 All (　　) pipes should be adequately lagged.

가. marine diesel oil　　　　나. fresh water

사. control air　　　　아. steam

287 (　　) means a member of the ship's crew other than the master or an officer.

가. Engineer officer　　　　나. Radio officer

사. Navigation officer　　　　아. Rating

288 Before the maintenance of hydraulic equipment is undertaken, all (　　) within the system should be released.

가. temperature　　　　나. seals

사. water　　　　아. pressure

해설 maintenance : 유지 / undertake : ~을 맡다. 책임지다. 착수하다.

289 Warmed up M/E (　　) four hours.

가. for　　　　나. during

사. with　　　　아. in

290 All electric cables run through water tight bulkheads must be (　　).

가. welded on both sides of the bulkhead

나. installed with water tight packing glands

사. grounded on both sides of the bulkhead

아. fitted with unions on each side of the bulkhead

해설 bulkhead : 격벽

291 (　　) of a space is the percentage of that space which can be occupied by water in accordance with the provision of the SOLAS '74 convention.

가. Permeability　　　　나. Availability

사. Probability　　　　아. Stability

해설 provision : 조항, 규정 / permeability : 침투성, 투과성 / availability : 유용성 / probability : 가망성, 확률 / stability : 안정성

292 Theoretically, an increase of the compression ratio of an engine raises its (　　) and lowers its (　　). However, an increase of the compression ratio increases the maximum gas pressure and the combustion temperature.

가. mechanical efficiency, fuel consumption

나. thermal efficiency, fuel consumption

사. mechanical efficiency, brake horse power

아. thermal efficiency, brake horse power

293 A heat engine is a machine for the conversion of heat energy into (　　), and its function must be to make the conversion as efficiently as possible.

가. reciprocating motion　　　　나. centrifugal force

사. mechanical energy 아. resistance

294 The ( ) is a heat exchanger which removes the latent heat from exhaust steam so that it condenses and can be pumped back into the boiler.

가. feed water heater 나. superheater

사. condenser 아. feed water cooler

295 ( ) means a coating, painting, surface treatment, surface, or device that is used on a ship to control or prevent attachment of unwanted organisms.

가. Anti-fouling system 나. Anti-heeling system

사. Anti-surging system 아. Anti-vibration system

해설 organisms : 유기체, 생물 / foul : 구역질나는, 구린내 나는

296 Reduced RPM of main engine and slowed down speed according to order from bridge. What does the underlined RPM stand for?

가. Revolutions Per Minute 나. Reverutions Per Minite

사. Revolutions Per Minite 아. Reverutions Per Minute

297 ( ) means a ship whose hull, machinery, equipment, or operational safety is substantially below the standards required by the relevant convention or whose crew is not in conformance with the safety manning document.

가. Substandard ship 나. Designated ship

사. Flag ship 아. Registered ship

298 Propeller blades to be polished and surveyed. Clearance between ( ) and tail shaft to be measured.

가. propeller 나. rudder

사. stern tube 아. intermediate shaft

299 To avoid excessive surge pressure when bunkering, you should (   ).

가. shut valves before topping off
나. close valves to another tank
사. close valves slowly
아. spill bunker overboard

300 The engineer of the watch is (   ) to absent himself from the engine room, during his watch, on any account.

가. usually
나. sometimes
사. used
아. never

301 In accordance with the SOLAS74/78 Convention as amended, the lifeboat engine shall be capable of operating when the (   ) is flooded up to the centerline of the crankshaft.

가. lifeboat
나. mother ship
사. generating engine
아. main engine

302 According to the MARPOL 73/78 Convention as amended, the (   ) shall be kept in such a place as to be readily available for (   ) at all reasonable times, and except in case of unmanned ships under tow, shall be kept on board the ship.

가. DoC, inspection
나. oil record book, inspection
사. DoC, protection
아. oil record book, protection

해설 tow : 끌려가는 배 / unmanned : 사람이 없는

303 "수동시동이 유효하다는 것을 입증 할 수 없는 한, 30분 이내에 추가하여 3회의 시동을 하기 위한 제2의 에너지원을 비치하여야 한다."를 영문으로 옮길 때 (   )에 알맞은 것은?

"A second source of energy shall be provided for (   ) within 30 minutes unless manual starting can (   ) to be effective."

가. an consecutive three starts, demonstrate
나. an consecutive three starts, be demonstrated

사. and additional three starts, demonstrate

아. an additional three starts, be demonstrated

해설 consecutive : 연속적인, 계속되는

304 다음 문장의 괄호 부분의 우리말 해석은?

> Before reaching a home port should be ready a complete (list of new stores) required.

가. 새로운 소모품 일람표　　　나. 식료품 소모량 일람표

사. 새로운 유류 소비량 일람표　　　아. 식료품 재고량 일람표

305 All engineer officers and engine room (　　) shall be aware of the serious effects of operational or accidental (　　) of the marine environment.

가. ratings, pollution　　　나. mates, condition

사. wipers, responsibility　　　아. oilers, circumstance

해설 be aware of : 깨닫다. 감지하다. 알아차리다. / accidental : 우발적인, 뜻하지 않은

306 Prior to taking over the engineering watch, relieving officers shall satisfy themselves regarding at least the following, that is the (　　) and special instruction of the chief engineer officer relating to the operation of the ship's system and machinery.

가. standing orders　　　나. operation

사. maintain　　　아. upkeep

해설 take over : ~을 이어받다. 인계받다. 접수하다. / instruction : 지시, 명령 / standing orders : 복무규정 / maintain : 계속하다. 유지하다.

307 "나머지는 취소할 수 밖에 없습니다."를 영문으로 옮길 때 (　　)에 알맞은 것은?

> "I (　　) no alternative to cancel the (　　) of our order."

가. have, surplus　　　나. have, remainder

사. take, surplus　　아. take, remainder

해설 alternative : 양자택일의, 선택의 / surplus : 나머지, 과잉, 필요이상의 잉여 / remainder : 나머지, 잉여, 소비한 나머지

308 A circuit that dose not provide a complete path for the flow of current is (　　).

가. a closed circuit　　나. a short circuit

사. an open circuit　　아. a grounded circuit

309 Fill blank(s) with suitable word(s).

> Procedure to be followed to prevent air and water pollution and measures to be taken in the event of spillage, including the need to promptly (　　) shore-based response personnel.

가. present　　나. notify

사. detect　　아. hear

해설 procedure : 진행, 절차 / spillage : ~을 엎지르다. 흘리다. / promptly : 기민하게, 즉시 / response : 응답 / personnel : 전직원, 요원 / notify : 알리다. / detect : 발견하다. 찾아내다.

310 Fill blank(s) with suitable word(s).

> This paragraph is Special Survey for Machinery. All air receivers and other pressure vessels for essential services together with their mountings and safety devices are to be opened up and examined (　　) and externally. If internal examination of them is not practicable, they are to be tested hydraulically to 1.5 times the (　　) pressure.

가. test hammer, gage　　나. eye, absolute

사. internally, working　　아. hydraulically, design

해설 paragraph : 항목 / air receiver : 에어탱크 / externally : 외부

311 Fill blank(s) with suitable word(s).

> Prior to taking over the engineering watch, relieving officers shall satisfy themselves regarding at least the following, that is, the nature of all work being performed on machinery and system, the personnel involved and potential (   ).

가. standing orders　　나. hazards

사. ability　　아. requirements

해설 prior to : ~에 앞서, 전에 / take over : 인계받다. / relieving officer : 교대자 / personnel : 직원들 / potential : 가능한, 잠재력이 있는 / hazard : 위험

312 Where main or auxiliary machinery including pressure vessels are subject to internal pressure and may be subject to dangerous (   ), means shall be provided where practicable to protect against such excessive pressure in accordance with the SOLAS 74/78 Convention.

가. overspeeding　　나. overpressure

사. breakdown　　아. high temperature

해설 subject to : 받아야하는, 필요하다. / convention : 협약

313 The jacket cooling water should be heated slowly, by circulating fresh water from the auxiliary generator discharge through M/E system. Where such a connection is not provided, (   ) may be arranged to percolate into the cooling water.

가. hot gas　　나. steam

사. hydraulic　　아. stream

314 This inspection was not a full survey and deficiencies listed may not be exhaustive. In the event of a detention, it is (   ) that full survey is carried out and all deficiencies are rectified before an application for reinspection is made.

가. contacted　　나. helped

사. recommended 아. operated

해설 survey : 검사 / deficiencies : 부족, 결핍, 결함 / exhaustive : 철저한, 남김 없는, 고갈하는 / detention : 억류, 지연, 감금 / rectify : 바로잡다. 고치다.

315 Where risk from ( ) of machinery exists, means shall be provided to ensure that the safe speed is not exceeded in accordance with the SOLAS 74/78 Convention.

가. overspeeding 나. overpressure

사. breakdown 아. high temperature

316 ( ) means a voyage by a ship entitled to fly the flag of one State to or from a port, shipyard, or offshore terminal under the jurisdiction of another State.

가. Coastal voyage 나. Domestic voyage

사. International voyage 아. Ocean going voyage

317 If the PSCO from general impressions or observations on board has clear grounds for believing that the ship might be substandard, the PSCO should proceed to a ( ) taking the following considerations into account.

가. initial inspection 나. periodical inspection

사. safety inspection 아. more detailed inspection

해설 PSC Port State Control : 외국선박검사. / observations : 관찰 / substandard : 표준이하의, 불충분한 / proceed : 착수하다. 계속하다.

318 Electric power is the rate of doing work by electricity. It is the (①) of voltage and current. The unit of power is watt or kilowatt. Kilo is a prefix which stand for (②).

가. ① ratio, ② 1000 나. ① product, ② 100

사. ① sum, ② 100 아. ① product, ② 1000

319 This paragraph is a part of 〈Request to apply repair〉. "Hereby I beg to apply to you that the under-detailed repairs of Engine Dep't on board the above named steamer will be completely (　　) by noon the 22nd instant, next Friday."

가. finished　　　　나. received

사. rung　　　　아. started

320 According to the SOLAS 74/78 Convention, (　　) of the ship is the length measured between perpendiculars taken at the extremities of the deepest subdivision load line.

가. draught　　　　나. breadth

사. length　　　　아. freeboard

해설 perpendiculars : 수직의, 직립한 / extremities : 끝, 선단 / subdivision : 세분 / draught : 흘수(=draft) / freeboard : 건현

321 According to the SOLAS 74/78 Convention as amended, float-free launching is that method of launching a (　　) whereby the craft is automatically released from a sinking ship and is ready for use.

가. rescue boat　　　　나. mobile supply vessel

사. survival craft　　　　아. davit

해설 launching : 발사, 진수

322 These are kind of the procedures before starting diesel engine ;

1. (　　　) all valve spindles well and ascertained no slightest leak of air, oil and water.
2. (　　　) at least one revolution with the turning gear, ascertained that everything was free.
3. (　　　) the fuel pump, and pumped some oil into the fuel valves.

가. Oiled, Cleaned, Renewed　　　　나. Oiled, Turned, Primed

사. Cleaned, Turned, Primed　　　　아. Cleaned, Turned, Renewed

해설 ascertained : ~을 확인하다. 규명하다. / not ~ in the slightest : 조금도 ~ 않다 / at least : 적어도

323 The mechanical efficiency of an engine is the ratio of the power actually delivered to the crank shaft called (①) to the power developed inside the cylinders called (②).

가. ① DHP, ② IHP　　나. ① IHP, ② BHP

사. ① BHP, ② IHP　　아. ① DHP, ② BHP

324 The life-raft and its fittings shall be so constructed as to enable it to be towed at a speed of (　　) knots in calm water when loaded with its full complement of persons and equipment and with one of its sea-anchors streamed according to SOLAS convention.

가. 1　　나. 3

사. 6　　아. 10

해설 enable : ~하는 것을 가능하게하다. / tow : (배)를 밧줄로 끌다.

325 What does the following mean?

> A condition found not to be in compliance with the requirements of the relevant convention.

가. clear grounds　　나. detention

사. inspection　　아. deficiency

해설 compliance : 응함, 따르기 / relevant : 관련된　convention : 협정, 관례 / detention : 구금, 구류 / inspection : 검사, 점검 / deficiency : 결함, 불충분

326 If there are clear grounds to believe that the ship is in violation of this Convention, a through inspection may be carried out (　　) account guidelines developed by the Organization.

가. taken into　　나. put on

사. putting on　　아. taking into

해설 in violation of : ~에 위반해서 / convention : 협약

327 Heat treatment consists of heating a metal alloy to the temperature below its (　　) and then cooling it in a particular manner. The result is some desired change in the material properties.

가. boiling point　　　　나. pour point

사. melting point　　　　아. ignition point

해설 heat treatment : 열처리 / alloy : 합금, 비금속 / boiling point : 비등점 / pour point : 유출점 / melting point : 융해점 / ignition point : 착화점

328 In a turbine, the steam passes through a series or succession of wheels holding (①). It passes through them one after the other. Each successive row reduces the (②) energy of the steam. The energy of the steam is (③) reduced as it passes through the successive rows.

가. ① stages, ② moving, ③ slowly

나. ① rotor ② stored ③ gradually

사. ① blades, ② kinetic, ③ progressively

아. ① disc, ② potential, ③ greatly

해설 succession : 연속적으로 / row : 줄, 열

329 We can use grease in low-speed bearings, but grease offers more (①) to the turning movement of the shaft. A lighter oil causes less (②), and so an oily lubricant is better for high-speed bearing.

가. ① viscosity, ② friction　　　　나. ① power, ② output

사. ① resistance, ② friction　　　　아. ① viscosity, ② resistance

330 Choose the most suitable one for the blanks.

> (　　) is the control of temperature and humidity in a space together with the circulation, filtering and refreshing of the air. (　　) is the circulation and refreshing of the air in a space without necessarily a change of temperature.

가. Ventilation, Air conditioning
나. Refrigeration, Ventilation
사. Air conditioning, Ventilation
아. Ventilation, Refrigeration

331 The ratio between the air-fuel mixture that enters the cylinders and the volume of the same mixture that could enter under standard conditions of temperature and pressure is called the (   ).

가. volumetric efficiency
나. volume ratio
사. volume rating
아. volumetric ratio

332 This paragraph is a part of "Articles Order". Please fill out the blank with suitable words. I have the pleasure of forwarding you an order of the undermentioned goods and shall be so much obliged if you send them on board M/V ZZZ immediately. You will be paid by our agents, Messrs. Busan & Co., in exchange for (   ) of goods.

가. order
나. receipt
사. contain
아. quantity

해설 paragraph : 단락 / Articles Order : 물품주문, 상품주문 / undermentioned : 하기의, 다음에 말하는 / oblige : ~에게 강요하다. 억지로 시키다. / agent : 대리점 / in exchange for : ~과 교환으로, ~의 대신으로

333 The safety officer should (   ) the master on all matter of safe working practice and assist him in the elimination of (   ) and injuries on board ship.

가. order, accidents
나. advise, theft
사. advise, accidents
아. order, theft

해설 elimination : 제거, 삭제, 배제

334 The contractor shall obtain the (   ) dry docking plan for the vessel's master and/or the owner's representative and (   ) the vessel on the keel and bilge block as to avoid the vessel's landing on previous block locations in so far as possible. Docking or undocking outside normal hours shall be for contractor's convenience and any additional charges shall be for contractor's account.

가. previous, position　　나. relevant, hold

사. important, inspect　　아. good, survey

해설 contractor : 계약자, 청부인 / representative : 대리인 / keel : 용골 / as to : ~에 관하여 / in so far as : ~하는 한 / convenience : 편리 / account : 보고(서) / previous : 앞의, 이전의 / relevant : 관련된, 관련 있는

335 The reason I pointed out the above deficiency were resulted from the aforementioned principle and also my professional judgements (　　) the SOLAS & TOKYO MOU.

가. of　　나. according to

사. on　　아. due to

해설 pointed out : ~을 지적하다. / deficiency : 부족, 결핍 / aforementioned 전술한, 전기한 / according to ~에 따라, ~에 의하면

336 (　　) means an inspection conducted when there are clear grounds for believing that the condition of the ship, its equipment, or its crew does not correspond substantially with particulars of the certificates.

가. Concentrate inspection　　나. More detailed inspection

사. Random Inspection　　아. Periodical inspection

해설 conducted : ~을 실시한다. 안내하다. / equipment : 설비 / substantially : 충분히, 대체적으로 / certificates : 증명서 / periodical : 주기

337 Fill blank(s) with suitable word(s).

> When megger-testing a new armature coil to ground, the reading under normal conditions should be at least (　　)[MΩ].

가. 1　　나. 10

사. 100　　아. 1000

338 (　　) provides a reservoir of relatively cool water giving the gravitational head necessary to displace the high temperature mixture of steam and water, with its much smaller density, from the heated tube surfaces.

가. Water drum　　나. Scotch boiler

사. Steam drum　　아. Evaporating tube

해설 relatively : 상대적으로, 비교적 / gravitational : 중력의 / density : 밀도

339 (　　) to be removed ashore, tips of two blades to be faired where distorted, and ground smooth. After proper fit of shaft taper in hub is witnessed, (　　) to be replaced in position ready for installation.

가. Propeller　　나. Intermediate shaft

사. Side-thruster　　아. Stern tube

해설 faired : ~반반하게 다듬다. / distorted : 일그러지게 하다. 뒤틀다. / witnessed : 입증하다. 보이다. 입회하다.

340 Immersion suit can be unpacked and donned without assistance within (　　), taking into account any associated clothing and a life jacket if the immersion suit is to be worn in conjunction with a life jacket according to SOLAS Convention.

가. 30[sec]　　나. 2[min]

사. 5[min]　　아. 10[min]

해설 immersion : 침투, 침수 / unpacked : ~의 내용물을 빼내다. 열다. / assistance : 보조 / taking into : ~을 고려하다. / associated : 교제하다. 연합 / conjunction : 결합, 연합, 합동

341 In accordance with the SOLAS Convention, the (　　) shall be capable of putting the rudder over from 35° on one side to 35° on the other side with ship at its deepest sea going draft and running ahead at maximum ahead service speed.

가. main and auxiliary steering gears　　나. main steering gear

사. auxiliary steering gear　　아. gyro system

342 The hand flare shall continue to burn after having been immersed for a period of 10 sec. under (　　) mm of water according to SOLAS Convention.

가. 100　　　　나. 500
사. 1000　　　　아. 5000

해설 flare : (불꽃이)타오르다. 확 타오르다. / immersed : 담그다.

343 Choose the most suitable one for the blanks.

> This paragraph is a part of "Request to offer Instruction Book." Please send Bulletin containing further information regarding the De Laval Oil Purifier, as (　　) below : # Purification of Diesel Lubricating Oil Name; AAA, Company; BBB, Address; CCC.

가. checked　　　　나. made
사. recommended　　　　아. operated

해설 bulletin : 게시, 고시, 공보, 편람 / purification : 정화, 청정

344 The (　　) battery is made up of a series of cells. One cell consists of a lead peroxide positive plate and a lead negative plate both immersed in a dilute sulfuric acid solution. The sulfuric acid is known as the electrolyte.

가. alkaline　　　　나. dry
사. lead　　　　아. Ni-Cd

해설 peroxide : 과산화물 / immersed : 담그다. / dilute sulfuric acid solution : 황산수용액(희석액)

345 Choose the most suitable one for the blanks.

> All waste removed from the ship as a result of this contract shall be disposed by (　　) according to the requirements of local and national laws of the country to which the contract is awarded.

가. captain　　　　나. chief engineer

사. contractor 아. owner's representative

해설 waste : 쓰레기 / contract : 협정, 협약 / disposed : 처리하다. / award : (상을)주다. ~을 개정하다.

346 Choose the most suitable one for the blanks.

( ) means a vessel capable of engaging in drilling operations for the exploration for or exploitation of resources beneath the sea-bed such as liquid or gaseous hydrocarbons, sulphur or salt.

가. Mobile offshore drilling unit 나. Bulk carrier
사. Oil tanker 아. Chemical tanker

해설 drilling operations : 시추 / exploration : 실지답사, 탐구 / exploration : 개발 / sea-bed : 해저

347 Choose the most suitable one for the blanks.

( ) means a document issued to a Company which complies with the requirements of ISM Code.

가. Safety Management Certificate 나. Document of Compliance
사. Safety Management System 아. ISPS Certificate

해설 complies : 따르다, 동의하다 / Compliance : 수락, 승낙

348 중요한 보조 기계 1개가 작동불능이 된 경우에도 추진 기관의 정상운전을 유지 또는 회복시킬 수 있는 수단을 강구하여야 한다.를 영문으로 옮길 때 ( )에 알맞은 것은?

( ) shall be provided whereby normal operation of propulsion machinery can be sustained or restored even though one of the essential auxiliaries becomes ( ).

가. Methods, inoperative 나. Means, inoperative
사. Methods, operative 아. Means, operative

해설 even though : 비록 ~일지라도 / Means : 수단

349 Choose the most suitable one for the blanks.

> (   ) means the gross tonnage calculated in accordance with the tonnage measurement regulation contained in Annex1 to the International Convention on the Tonnage Measurement of Ships, 1969, or any successor Convention.

가. Net tonnage   나. Gross tonnage
사. Displacement tonnage   아. Dead Weight tonnage

해설 in accordance with : ~에 따라서, ~과 일치하여 / Annex1 : 부록1

350 Choose the most suitable one for the blanks.

> The (   ) is a forum in which Member States can express their views reflecting state or international shipping practice.

가. ILO   나. WHO
사. IMO   아. IOC

해설 reflecting : 반사하다, 반영하다

351 The ratio between the air-fuel mixture that enters the cylinders and the volume of the same mixture that could enter under standard conditions of temperature and pressure is called the (   ).

가. volumetric efficiency   나. volume ratio
사. volume rating   아. volumetric ratio

352 (   ) are used to ensure that ships do not discharge oil when pumping out bilges, oil tanks or any oil contaminated space.
International legislation relating to oil pollution is becoming more and more stringent in the limits set for oil discharge.

가. Oil spill removement   나. Waste oil incinerator
사. Oily water separator   아. Oil pollution simulator

해설 legislation : 법률제정, 입법, 법규 / stringent : 엄한, 엄격한

353 6월 28일에 수량, 품질, 가격 및 품명에 관한 모든 명세를 부기하여 선용 소모품을 귀사에 주문하였습니다.

I forwarded you on the 28th of June an order for (　　), with all particulars as to quantity, (　　), price and name.

가. ship's spare, quality
나. ship's spare, specification
사. ship's consumables, quality
아. ship's consumables, specification

해설 as to : ~에 관하여 / quantity : 양

354 Before using a megger it should be tested. This is done by placing the test leads together and turning the crank. A perfect reading, indicating that the leads and megger are in good condition, reads (　　).

가. zero
나. 100
사. 220
아. 440

355 This paragraph is a part of "Protection of the Marine Environment." All engineer officers and engine room ratings shall (　　) all possible precautions to (　　) such pollution, particularly within framework of relevant international and port regulations.

가. disregard, prevent
나. take, prevent
사. disregard, protect
아. take, protect

해설 precautions : 주의 / framework : 구조물 / disregard : ~을 무시하다. 소홀히 하다. / prevent : 예방하다. / protect : 보호하다.

356 If there are clear grounds to believe that the ship is (　　) violation of this Convention, a through inspection may be carried out taking into account guidelines developed by the Organization.

가. on
나. to

사. in　　　　아. with

해설 in violation of : ~에 위반하여

357 In accordance with the MARPOL 73/78 Convention as amended, the oil filtering system shall be equipped with (　　) which will ensure that the discharge is automatically stopped when the oil content of the effluent exceeds 15 parts per million.

가. a discharge monitor　　　　나. a by-pass device

사. a stopping device　　　　아. a recording device

해설 in accordance with : ~에 따라서 / discharge : 배출

358 Claw couplings or clutch couplings are used for (　　) or (　　) shafts when the change is to be made often.

가. connecting - disconnecting　　　　나. driving - undriving

사. sliding - intersecting　　　　아. connecting - rotating

해설 claw : 서로 맞물리는 커플링

359 During the period of maneuvering engine, careful attention must continue to be given to the jacket water temperatures, the sea water supply to the fresh water cooler being severly throttled or, completely (　　) for ensuring a satisfactory temperature level.

가. opened　　　　나. increased

사. shut-off　　　　아. repaired

해설 maneuvering : 수동 / severly : 엄격하게, 심하게 / ensuring : 안전하게하다, 책임지다, 지키다

360 The "officer in charge of the engineering watch" is the chief engineer officer's representative and is responsible for inspection, (　　) and testing, as required, of all machinery and equipment under the responsibility of the engineering watch.

가. maintain　　　　나. operation

사. survey　　아. upkeep

해설 representative : 대리인 / responsible : 책임이 있는 / responsibility : 책임

361 According to the Annex IV of the MARPOL 73/78 Convention as amended, (　　) means drainage and other wastes from any form of toilets, urinals, and WC scuppers.

가. garbage　　나. sewage

사. noxious liquid　　아. sludge

362 In accordance with the SOLAS 74/78 Convention as amended, remote control of the propulsion machinery shall be possible only from (　　) at a time.

가. one location

나. two locations

사. the engine control room

아. the navigation bridge and engine control room

해설 in accordance with : ~에 의하면, 따라서 / propulsion machinery : 추진장치 / at a time : 한번에, 동시에

363 Two or more prime movers (diesels) are electrically interconnected and driving a common shaft and load. What is this called?

가. Double acting　　나. Twin engine

사. Parallel　　아. Speed droop

364 The (　　) slip is the ratio of the difference between the speed of the propeller and the speed of the ship over the ground in still water to the speed of the propeller.

가. true　　나. effective

사. apparent　　아. propeller

해설 apparent slip : 겉보기 슬립

365 When the ship is at sea, the principal particulars to be entered in the log book are as follows : The (　　) of arrival and departure of the ship, the (　　) and details of the starting and stopping of the engines, and the (　　) and reason for reducing the speed.

가. time　　나. place
사. condition　　아. reason

366 (　　) Code means the International Management Code for the Safe Operation of Ships and for Pollution Prevention adopted by the IMO.

가. ISPS　　나. STCW
사. LSA　　아. ISM

해설 adopted : 채택된

367 (　　) is the machinery, rudder actuators, steering gear power units, if any, and ancillary equipment and the means of applying torque to the rudder stock necessary for effecting movement of the rudder for the purpose of steering the ship under normal service conditions.

가. Main steering gear　　나. Steering gear control system
사. Steering gear indicating system　　아. Rudder keeping system

해설 ancillary : 보조적인, 부속적인, 부수적인

368 (　　) are those spaces in which the ship's radio or main navigating equipment or the emergency source of power is located or where the fire recording of fire control equipment is centralized in accordance with the provision of the SOLAS convention.

가. Control rooms　　나. Bridges
사. Machinery spaces　　아. Control stations

해설 centralize : ~을 집중하다. 중앙화하다. / provision : 조항, 규정

369 When CO_2 is used as the extinguishing medium for spaces containing boilers or

internal combustion type machinery the fixed piping system shall be such that ( ) of the gas can be discharged into the space within ( ) according to SOLAS.

가. 85% – 3min. 나. 85% – 2min.

사. 90% – 3min. 아. 90% – 2min.

370 The PSCO may ensure that ( ) are exhibited in conspicuous places throughout the ship, including the navigational bridge, the engine room and crew accommodation spaces.

가. crew lists 나. muster lists

사. spare lists 아. provision lists

해설 exhibited : 전시하다 / conspicuous : 잘 보이는, 눈에 잘 띄는 / muster list : 점검 목록

371 "외부로부터 아무런 도움 없이 데드쉽 상태에서 기관을 작동시킬 수 있는 수단을 강구하여야 한다."를 영문으로 옮길 때 ( ) 속에 적합한 것은?

> Means shall be provided to ensure that the machinery can ( ) from the dead ship condition without ( ).

가. bring into operation, internal aid.

나. be brought into operation, internal aid

사. bring into operation, external aid

아. be brought into operation, external aid

372 Main and emergency switchboards, section panels, and sub-distribution fuse panels are to be examined, and ( ) protective devices and fuses to be inspected to verify that they provide suitable protection for their respective circuits.

가. relay 나. switchboard

사. overcurrent 아. circuit

해설 sub-distribution : 분전반 / verify : ~을 검정하다. 입증하다. 확인하다.

373 The temperature to which the fuel must be heated to allow flammable vapors to form on its surface. These vapors can be ignited by an open flame held above the surface. The importance of (　　) is in storage safety, not performance in an engine.

가. pour point　　나. flash point
사. ignition point　　아. combustion point

해설 performance : 성능 / pour point : 유동점 / flash point : 인화점 / ignition point : 착화점

374 (　　) may be attached at the "free" end of the engine, if necessary, so as to avoid inadmissibly heavy vibrations of the crankshaft when passing through the critical speed ranges.

가. A flywheel　　나. A thrust bearing
사. A tacho-meter　　아. A vibration damper

해설 inadmissibly : 허용할 수 없는

375 A tanker of ten years of age and over shall undergo a minimum of (　　) survey during the period of validity of its cargo ship safety construction certificate in accordance with the relevant provisions of the SOLAS '74 convention.

가. one annual　　나. one intermediate
사. one unscheduled　　아. one special

해설 intermediate : 임시

376 All of the load developed in the cylinder passes through (　　). It is the only link between the piston and connecting rod. In larger engines the piston pins are sometimes locked in the piston at both ends.

가. the piston pin　　나. the piston crown
사. the crosshead　　아. the crankshaft

377 "A급 구획은 1시간의 표준화재 시험이 종료할 때까지 연기 및 화염의 통과를 방지 할 수 있도록 제조되어야 한다."를 영문으로 옮길 때 (　　) 속에 적합한 것은?

> The class A (   ) shall be so constructed as to be capable of (   ) the passage of smoke and flame to the end of the one-hour standard fire test.

가. bulkhead, prevention　　나. bulkhead, preventing

사. division, prevention　　아. division, preventing

해설 bulkhead : 격벽

378 It is essential (①) the inlet and exhaust valves to open and close exactly at the appropriate moment in relation to the position of the (②). Therefore they are actuated by a camshaft running in phase with the (③).

가. ① to, ② valve, ③ crankshaft　　나. ① in, ② piston, ③ engine

사. ① for, ② piston, ③ crankshaft　　아. ① with, ② valve, ③ piston

379 The purpose of (   ) is to scrape off most of the lubricating oil splashed upward by the crank shaft and connecting rod in order to reduce the amount of oil carried upward and burnt in the combustion chamber.

가. top ring　　나. compression rings

사. oil rings　　아. "O" ring

해설 splash : ~을 튀기다.

380 "본인은 거룻배 및 예인선의 소유자가 본건 피해의 책임을 져야 한다고 믿고 있습니다."

> "I (   ) the owners of the lighter and tug boat (   ) the damage sustained."

가. exempt, responsible for　　나. exempt, responsible of

사. hold, responsible for　　아. hold, responsible of

381 "다음 품목의 주문서를 보내오니 3월5일 까지 본선으로 납품하여 주시면 고맙겠습니다."를 영문으로 옮길 때 (   )에 알맞은 것은?

"We (　　) sending you an order of the undermentioned goods, and shall be (　　) you send them on board by March 5th."

가. are pleased to, glad if　　　　나. are pleased, glad that

사. have the pleasure of, glad if　　　　아. have the pleasure of, glad that

382 Which of the following is correct in docking order?

① Arrived at the dry dock entrance and proceeded to haul into the dock.
② Commenced to let water into the dry dock and vessel floated.
③ A dockmaster came on board and the ship was towed by two tugs toward the dock.
④ Commenced to pump out water of the dock.

가. ① - ② - ③ - ④　　　　나. ① - ③ - ④ - ②

사. ① - ④ - ② - ③　　　　아. ① - ② - ④ - ③

383 Officers in charge of an engineering watch, as defined in the STCW Code and under the direction of the chief engineer officer, when required, shall be physically (　　) in the machinery space during their periods of responsibility.

가. present　　　　나. watch

사. stand　　　　아. keep

384 Free-fall (　　) is the rate of change of velocity experienced by the occupants during launching of a free-fall lifeboat.

가. height　　　　나. weight

사. acceleration　　　　아. velocity

해설 free-fall : 자유낙하 / occupants : 거주자 / launching : (보트를)물에 띄우다. 진수시키다. / acceleration : 가속

385 Length of the ship is the length measured between ( ) taken at the extremities of the deepest subdivison load line in accordance with the provision of the SOLAS'74 convention.

가. perpendiculars 나. frames
사. draughts 아. freeboard

해설 extremities : 끝, 선단 / subdivison : 세분 / load line=water line / perpendiculars : 수직의

386 "비상전원은 최상층 전통갑판위 및 충돌격벽의 후방에 위치하여야 한다."를 영문으로 옮길 때 ( )에 알맞은 것은?

"The emergency source of electrical power shall be located above ( ) and backward of ( )."

가. the uppermost continuous deck, the collision bulkhead
나. the uppermost open deck, the collision bulkhead
사. the uppermost continuous deck, the watertight bulkhead
아. the uppermost open deck, the watertight bulkhead

387 If carbon dioxide has been discharged to extinguish or prevent a fire, the ( ) content will be reduced in the affected space.

가. hydrocarbon 나. hydrogen
사. carbon dioxide 아. oxygen

388 Power factor, by definition, is the ( ) of the active power to the apparent power.

가. product 나. minus
사. ratio 아. sum

해설 definition : 정의 / apparent : 눈에 보이는, 분명한 / apparent power : 겉보기 힘

389 According to the MARPOL 73/78 Convention as amended, ( ) of discharge of oil content means the rate of discharge of oil in litres per hour at any instant divided by the speed of the ship in knots at the same instant.

가. instantaneous quantity　　나. instantaneous rate

사. persistant quantity　　아. persistant rate

해설 instantaneous rate of discharge : 순간배출율

390 Adjusted ( ) of fuel oil valve to 300 [kg/cm^2].

가. injection pressure　　나. atomization

사. viscosity　　아. penetration

391 Wiper, Kim ( ) duty, being sick.

가. on　　나. off

사. out　　아. in

392 Every oil tanker of 150 tons gross tonnage and above and every ship of ( ) tons gross tonnage and above other than an oil tanker shall be provided with an oil record book as part of the ship's offical logbook according to MARPOL 73/78 Convention.

가. 200　　나. 300

사. 400　　아. 500

393 "해상인명안전협약의 제 II-1장 26규칙에 의거 선박용 기관은 선내 승조원에 대한 위험을 최소한으로 할 수 있도록 설치 되어야 한다."를 영문으로 옮길 때 ( )에 알맞은 것은?

"The machinery on board ( ) be so installed as to reduce to a minimum any danger to persons on board ( ) the regulation 26 of chapter II-1 in the SOLAS Convention."

가. shall, in accordance to　　나. shall, in accordance with

사. will, in accordance with　　아. will, in accordance to

해설 in accordance with : ~에 따라서, ~와 일치하여

394 (　　) are used to ensure that ships do not discharge oil when pumping out bilges, oil tanks or any oil contaminated space. International legislation relating to oil pollution is becoming more and more stringent in the limits set for oil discharge.

가. Oil spill removement　　나. Waste oil incinerator
사. Bilge separator　　아. Oil pollution simulator

해설 ensure : ~을 책임지다. 보증하다.

395 This fresh water generator is a vacuum evaporation distiller, normally using (　　) from the fresh cooling water of a diesel engine as (　　).

가. the waste heat, heating medium　　나. the waste heat, cooling medium
사. the latent heat, heating medium　　아. the latent heat, cooling medium

396 The Chief Engineer is responsible for the ship's engines. He is (　　) charge of the Engine Department.

가. at　　나. to
사. off　　아. in

해설 in charge of : ~을 맡은

397 The (　　) limiter defines a maximum amount of fuel to be injected according to actual engine speed. This is to ensure that the engine (　　) does not exceed recommended levels.

가. torque　　나. overspeed
사. start　　아. scavenge air pressure

398 Reduce the engine load to (　　) of normal. Then, by means of the thermostatic valve in the steam system, or by manual control of the viscosity regulator, the diesel oil is heated to maximum 60~80℃, Regulate the preheating so as to give a temperature rise of about (　　)℃ per minute.

가. 1/2, 2　　나. 2/3, 5

사. 3/4, 2　　　　아. full, 5

해설 so as to~ : ~하도록

399 (　　) is the difference in tonnes between the displacement of a ship in water of a specific gravity of 1.025 at the load waterline corresponding to the assigned summer freeboard and the lightweight of the ship.

가. Deadweight　　　　나. Gross tonnage

사. Net tonnage　　　　아. Displacement weight

해설 deadweight : 중량 / gross tonnage : 총톤수 / net tonnage : 순톤수

400 Use gloves when removing O-rings and other rubber/plastic-based sealing materials which have been subjected to abnormally high temperatures. These materials may have a (　　) effect when being touched directly. The gloves should be made of neoprene or PVC.

가. caustic　　　　나. e

사. clogging　　　　아. side

401 "우선 입거기간은 14일 내지 15일 동안으로 예약 하겠습니다."를 영문으로 옮길 때 (　　)에 알맞은 것은?

> "The dry dock should (　　) initially (　　) 14 to 15 days."

가. be booked, for the period of　　　　나. book, for the period of

사. be booked, by the period of　　　　아. book, by the period of

402 With stopped engine, blow-by can be located by inspecting the condition of the piston rings, through the (　　). Piston and cylinder become black in the area of blow-by.

가. fuel valve hole　　　　나. cylinder lubricator hole

사. scavenge air ports　　　　아. test cock

403 The most important part of the frame of any ship is the bottom centerline, known as the (   ).

가. rake
나. superstructure
사. hull
아. keel

해설 rake : 경사 / hull : 선체 / keel : 용골

404 (   ) is inserted in the liquid line between receiver and evaporator for removing moisture in the refrigerant.

가. Dehydrator
나. Trap
사. Condenser
아. Strainer

해설 Dehydrator : 건조기, 건조제

405 In accordance with the SOLAS 74/78 Convention as amended, (   ) means a plan developed to ensure the application of measures on board the ship designed to protect persons on board, cargo, cargo transport units, ship's stores or the ship from the risks of a security incident.

가. garbage management plan
나. port facility security plan
사. shipboard oil pollution emergency plan
아. ship security plan

406 In accordance with the 1978 STCW 78 Convention as amended, all persons who are assigned duty as officer in charge of a watch or as a rating forming part of a watch shall be provided a minimum of (   ) hours of rest in any (   ) hour period.

가. 10, 24
나. 8, 24
사. 10, 12
아. 8, 12

407 (   ) are those used for public spaces, corridors, cabins, offices, pantries and lockers and similar spaces.

가. Control spaces
나. Accommodation spaces
사. Public spaces
아. Service spaces

해설 corridors : 복도

408 The (　　) is found in a natural state beneath the earth's surface. It was formed over long periods of time by decayed organic matter, subjected to the high temperatures and pressures found within the earth's crust.

가. diesel oil
나. crude oil
사. gasoline
아. kerosene

409 In accordance with the SOLAS 74/78 Convention as amended, (　　) power units shall be arranged to restart automatically when power is restored after a power failure.

가. main and auxiliary steering gear
나. main steering gear
사. auxiliary steering gear
아. emergency

410 According to the MARPOL 73/78 Convention as amended, product carrier means an oil tanker engaged in the trade of carrying oil (　　) crude oil.

가. including
나. such as
사. for instance
아. other than

해설 product carrier : 석유제품 / such as : ~과 같은 / for instance : 예를 들어 이를테면 / other than ~ : ~이외의, ~을 제외한

411 According to the SOLAS 74/78 Convention as amended, non-combustible material is a material which neither burns nor (　　) flammable vapours in sufficient quantity for self-ignition when heated to approximately 750℃.

가. gives off
나. gives over
사. puts over
아. puts off

해설 gives off : 방출하다. 발하다. / gives over : 그만두다. / puts over : 호평을 받게 하다. / puts off :

제거하다. 옷을 벗다.

412 "Faced up spare air starting valves for M/E." 밑줄친 부분과 같은 뜻을 가진 것은?

가. Checked　　나. Found

사. Lapped　　아. Welded

413 Adjusted injection (　) to 300[kg/cm^2].

가. temperature　　나. liquid

사. pressure　　아. level

414 "Adjusted regulating handle of fuel oil to 130(rpm)."의 문장에서 괄호의 "rpm" 약자는?

가. revolutions per minute　　나. revolutions part minute

사. revolutions per month　　아. rotor per month

415 Inspected all working parts in crankcase, found one crank pin bolt loosened, and (　).

가. renewed it　　나. took out it

사. retightened it up　　아. retightened it down

416 When is the most critical time of bunkering in regards to oil pollution?

가. lining up the system　　나. topping off

사. disconnecting hoses and joints　　아. connecting hoses and joints

417 For practical purposes (　) is determined by noting the number of seconds required for a given quantity at a particular temperature to flow through a standard orifice.

가. density　　나. viscosity

사. flow ratio　　아. pour point

418 Main bearings may be adjusted either by removing (　　) or grinding or filing the faces of the bearing caps.

가. metal　　나. bearing shell

사. bearing cover　　아. shims

419 Where a centrifuge is arranged to separate two liquids, it is known a(an) (　　). Where a centrifuge is arranged to separate impurities and small amounts of water from oil it is known as a(an) (　　).

가. clarifier, purifier　　나. strainer, filter

사. purifier, clarifier　　아. filter, strainer

420 The oil record book shall be kept in such a place as to be readily available for inspection at all reasonable times and preserved for a period (　　) years after the last entry has been made.

가. two　　나. three

사. four　　아. five

421 The ship sank a mile (　　) shore.

가. away　　나. on

사. off　　아. at

422 What does IMO stand for?

가. Inter-governmental Marine Organization

나. Inter-governmental Maritime Organization

사. International Merchant Marine Organization

아. International Maritime Organization

423 "Stopped cargo work owing to (　　) trouble." 에서 (　　) 속에 적합한 것은?

가. crane 나. windlass

사. anchor 아. telegraph

424 The refrigerant passes through the (①) in which the pressure falls at constant enthalpy. As the pressure falls some of the liquid flashes off, taking its (②) from the surrounding liquid, so causing a fall in temperature.

가. ① evaporating tube, ② liquid heat of evaporation

나. ① solenoid valve, ② entropy of refrigerant

사. ① condenser, ② sensible heat of refrigerant

아. ① throttling valve, ② enthalpy of evaporation

425 A pipe with a number of inlets or outlets from the several cylinders of an engine is ( ).

가. machinery 나. plant

사. manifold 아. unit

426 The branch of mechanics which treats of the stress and deformation of elastic materials is ( ).

가. thermodynamics 나. hydrodynamics

사. solid mechanics 아. fluid mechanics

427 The defining feature of a/an ( ) is that useful work is performed by the expanding hot gases acting directly to cause movement.

가. external combustion engine 나. marine boiler

사. steam turbine engine 아. internal combustion engine

428 During shop trial and sea trial, we recommend using a cylinder oil with a high detergency level. Use a "total base number" (TBN) of 70 as a 70 TBN oil will normally give good results. Use higher TBN oils in the event of high ( ) content in the fuel oil.

가. ash　　나. salt

사. sulphur　　아. water

해설 detergency : 세정력, 정화 / total base number(TBN) : 총알칼리가

429 Before engaging the turning gear, check that the (　　) supply is shut off, and that the (　　) are open.

가. lubricating oil, exhaust valves　　나. cooling fresh water, fuel valves

사. starting air, indicator cocks　　아. fuel oil, intake valves

430 Between the extremes of conductors and insulators are a group of elements known as (　　).

가. semiconductors　　나. metals

사. plastics　　아. electricity

해설 semiconductors : 반도체

431 The detector will give alarm and slow-down at a mist (　　) which is only a fraction of the lower explosion limit, LEL, to gain time to stop the engine before ignition of the oil mist can take place.

가. combustion　　나. concentration

사. evaporation　　아. condensation

432 If an empty tank has been left sealed for a long time, the oxygen content of the air in the tank (　　).

가. will be increased　　나. will be reduced

사. can not be changed　　아. may be increased or reduced

433 There are two types of (　　) in general use; the displacement or reciprocation type,

and the centrifugal type.

가. pumps
나. motors
사. engine
아. valves

434 Adjusted regulating handle of ( ) to average RPM 150.

가. F.O.
나. L.O.
사. F.W.
아. S.W.

435 "선박과 선박의 설비"를 영문으로 표기할 때 ( )에 알맞은 것은?

"Ship and ( ) is fittings."

가. my
나. his
사. her
아. she

436 The ( ) is the interval between injector opening and the start of ignition. During this period and the first droplets to enter the cylinder are heated by the surrounding charge of compressed air, begin to vaporize, and finally ignite.

가. rapid combustion period
나. steady burning period
사. after burning period
아. ignition delay period

437 If there are ( ) grounds to believe that the ship is in violation of this Convention, a through inspection may be carried out taking into account guidelines developed by the Organization.

가. clear
나. back
사. safe
아. doubtful

438 Every lifebuoy shall have an outer diameter of not more than ( ) mm and an inner diameter of not less than ( ) mm.

가. 600, 200
나. 800, 400

사. 200, 600

아. 400, 800

439 "주기의 연료 펌프를 개방 점검할 것"을 영문으로 옮길 때 (　　)에 알맞은 것은?

"Fuel pumps of main engine (　　)."

가. overhauled

나. to overhauled

사. to be overhauled

아. being overhaul

440 Depending upon the viscosity/temperature relationship and the viscosity index of the fuel oil, an outlet temperature of up to 150℃ will be necessary. This is illustrated in the diagram which indicates the expected preheating temperature as a function of the fuel oil viscosity. Recommended viscosity meter setting is (　　) cSt.

가. 0 - 5

나. 10 - 15

사. 20 - 30

아. 40 - 50

441 Mr. Han, 2nd engineer was sent to hospital (　　) of appendicitis.

가. due

나. as

사. because

아. owing

442 If the alarm on the fixed CO_2 system sounds in the E/R, you should (　　).

가. leave the space immediately

나. start the fire pump

사. make sure CO_2 starts flowing

아. close watertight door

443 (　　　) means an observed situation where objective evidence indicates the non-fulfilment of a specified requirement.

가. Observation

나. Objective evidence

사. Non-conformity

아. Administration

444 "2등 기관사가 연료유 계량에 종사했다."를 영문으로 옮길 때 (　　) 속에 적합한 말은?

> "2nd engineer engaged in (　　) F.O."

가. receiving　　나. repairing

사. tallying　　아. renewing

445 The crankshaft is of alloyed steel forging. To give better dynamic balancing, each (　　) is fitted with a counter-weight.

가. lever　　나. web

사. journal　　아. crank pin

446 Combustion knock in a diesel engine (　　).

가. is not related to ignition delay

나. is not harmful to the engine

사. will take place at the process of uncontrolled combustion period

아. is caused by turbulence and too many lighter molecules in the fuel

447 Every automated system should have a (　　) manually controlled.

가. demand display

나. feed back

사. back-up system

아. monitoring system

448 "중요한 보조기계 1개가 작동불능이 된 경우에도 추진기관의 정상운전을 유지 또는 회복시킬 수 있는 수단을 강구하여야 한다."를 영문으로 옮길 때 (　　)에 알맞은 것은?

> "(　　) shall be provided whereby normal operation of propulsion machinery can be sustained or restored even though one of the essential auxiliaries becomes (　　)."

가. Methods, inoperative

나. Means, inoperative

사. Methods, operative

아. Means, operative

449 "정지! 기름이 조금 새고 있습니다."를 영문으로 옮길 때 (　　) 속에 적합한 말은?

> "Stop! Oil is (　　) a little."

가. resisting　　나. entering

사. spilling　　아. charging

450 "불연성 재료라 함은 섭씨 750℃ 정도로 가열한 때에 연소하지 않으며, 자연 발화에 충분한 양의 인화성 증기를 발생하지 않는 재료를 말한다."를 영문으로 옮길 때 (　　)에 알맞은 것은?

> "Non-combustible material is a material which neither burns nor (　　) flammable vapours in sufficient quantity for (　　) when heated to approximately 750℃."

가. gives out, self-ignition

나. gives off, self-ignition

사. gives out, spark-ignition

사. gives off, spark-ignition

451 "Removed the starting pipes and <u>blinded off</u> the joints." 밑줄 친 부분과 같은 뜻을 가진 것은?

가. cleaned　　나. concealed

사. blew off　　아. blocked with stopper

452 Keep clear of space (　　) crane with load.

가. on　　나. by
사. up　　아. below

453 The (　　) of winch to be checked for smooth turning and greased at nipples.

가. sheaves　　나. lever pin
사. wire　　아. jib

454 "당직 중에 피스톤의 냉각유 출구온도를 65℃로 유지하였음"을 영문으로 표기할 때 (　　)에 알맞은 것은?

> "Kept piston cooling oil (　　) temp. to 65℃ during watch."

가. inlet　　나. onlet
사. outlet　　아. door

455 "양묘기는 선수에 위치한다."를 영문으로 옮길 때 (　　) 속에 적합한 용어는?

> "The (　　) is located in the bow of a ship."

가. winch　　나. windlass
사. hauler　　아. anchor

456 다음 단어들이 잘못 짝지어진 것은?

가. IMS Code - International Safety Management Code
나. ISPS Code - International Security and Pollution Survey Code
사. PSC - Port State Control
아. IOPP Cert. - International Oil Pollution Prevention Certificate

해설 ISPS Code란 "International Ship and port Facility Security Code"의 약자로 "국제 선박 및 항만시설 보안 코드"를 말하며, 2001. 9. 11 테러 이후 LNG 및 LPG 등을 포함한 해상화물 운송 선박 및 항만시설에 대한 해상 테러 가능성을 대비하기 위한 해상분야 보안 강화 규정을 SOLAS에 추가하기 위함.

457 다음 영문의 (　　) 속에 들어가야 할 가장 적당한 어휘는?

"The oil record book shall be completed on each occasion, on a (　　) basis."

가. drum - to - drum　　나. ship - to - ship

사. tank - to - tank　　아. bulk - to - bulk

정답

1	가	2	가	3	가	4	나	5	나	6	나	7	가	8	아	9	가	10	나
11	가	12	가	13	가	14	가	15	나	16	나	17	나	18	사	19	사	20	사
21	아	22	아	23	아	24	가	25	사	26	사	27	사	28	나	29	사	30	아
31	나	32	사	33	아	34	아	35	나	36	나	37	사	38	가	39	아	40	아
41	나	42	가	43	사	44	아	45	나	46	나	47	나	48	나	49	아	50	사
51	사	52	가	53	가	54	사	55	가	56	사	57	사	58	가	59	아	60	나
61	사	62	사	63	가	64	가	65	아	66	사	67	아	68	나	69	가	70	사
71	가	72	아	73	가	74	가	75	사	76	사	77	사	78	사	79	가	80	나
81	가	82	아	83	나	84	나	85	가	86	가	87	아	88	가	89	사	90	아
91	사	92	나	93	나	94	아	95	나	96	가	97	아	98	사	99	아	100	나
101	아	102	사	103	사	104	나	105	아	106	가	107	나	108	나	109	나	110	아
111	나	112	나	113	가	114	나	115	가	116	가	117	사	118	사	119	나	120	나
121	사	122	가	123	가	124	사	125	나	126	나	127	나	128	나	129	아	130	나
131	사	132	가	133	아	134	나	135	가	136	나	137	가	138	나	139	나	140	가
141	사	142	가	143	사	144	아	145	나	146	나	147	나	148	나	149	가	150	사
151	나	152	사	153	나	154	가	155	사	156	나	157	사	158	사	159	사	160	가
161	사	162	나	163	가	164	나	165	사	166	나	167	사	168	가	169	사	170	가
171	사	172	가	173	나	174	나	175	나	176	나	177	아	178	가	179	가	180	가
181	가	182	사	183	사	184	나	185	가	186	아	187	가	188	사	189	사	190	사
191	나	192	나	193	가	194	나	195	아	196	사	197	사	198	아	199	사	200	가
201	나	202	사	203	나	204	사	205	가	206	나	207	가	208	아	209	나	210	가
211	아	212	가	213	나	214	사	215	사	216	아	217	가	218	아	219	사	220	아
221	가	222	가	223	아	224	나	225	아	226	아	227	아	228	사	229	가	230	나
231	나	232	아	233	사	234	가	235	가	236	사	237	가	238	사	239	아	240	가
241	사	242	가	243	사	244	아	245	아	246	사	247	아	248	나	249	나	250	나
251	가	252	나	253	아	254	나	255	나	256	아	257	아	258	가	259	가	260	사
261	사	262	아	263	나	264	나	265	사	266	사	267	가	268	가	269	가	270	나
271	나	272	가	273	사	274	가	275	가	276	사	277	사	278	나	279	나	280	사
281	가	282	나	283	아	284	나	285	가	286	아	287	아	288	아	289	가	290	나
291	가	292	나	293	사	294	사	295	가	296	가	297	가	298	사	299	사	300	아
301	가	302	나	303	아	304	가	305	가	306	가	307	나	308	사	309	나	310	사
311	나	312	나	313	나	314	사	315	가	316	사	317	아	318	아	319	가	320	사
321	사	322	나	323	사	324	나	325	아	326	아	327	사	328	사	329	사	330	사
331	가	332	나	333	사	334	가	335	나	336	나	337	가	338	사	339	가	340	나
341	나	342	가	343	가	344	사	345	사	346	가	347	나	348	나	349	나	350	사
351	가	352	사	353	사	354	가	355	나	356	사	357	사	358	가	359	사	360	나
361	나	362	가	363	사	364	사	365	가	366	아	367	가	368	아	369	나	370	나
371	아	372	사	373	나	374	아	375	나	376	가	377	아	378	사	379	사	380	사
381	사	382	나	383	가	384	사	385	가	386	가	387	아	388	사	389	나	390	가
391	나	392	사	393	나	394	사	395	가	396	아	397	가	398	사	399	가	400	가
401	가	402	사	403	아	404	가	405	아	406	가	407	나	408	나	409	가	410	아
411	가	412	사	413	사	414	가	415	사	416	나	417	나	418	아	419	사	420	나
421	사	422	아	423	가	424	아	425	사	426	사	427	아	428	사	429	사	430	가
431	나	432	나	433	가	434	가	435	사	436	아	437	가	438	나	439	사	440	나
441	사	442	가	443	사	444	사	445	나	446	사	447	사	448	나	449	사	450	나
451	아	452	아	453	가	454	사	455	나	457	나	457	사						

참 고 문 헌

1. 교육과학기술부, 해사영어, 두산동아(주), 2010
2. 해문도서편찬회, 최신영문기관일지, 해문출판사, 1984

■ **저자약력**

장세호 / 부경대학교 기관학과 졸업(공학사)
부경대학교 대학원 공학석사
부경대학교 대학원 공학박사
부경대학교 실습선 기관장 역임
(현) 강원도립대학교 해양경찰과 교수

선박기관영어

인쇄 2018년 12월 24일
발행 2018년 12월 29일

저 자 장세호 저
발행인 임영철
발행처 도서출판 신화전산기획
주소 / 서울특별시 성동구 마조로 3나길 28
전화 / (02) 833-5145~6
팩스 / (02) 848-2498
등록 / 301-2010-034
홈페이지 / www.shj.co.kr

값 16,000원 ISBN 978-89-5534-228-4 93550